Social Skills and Work

Other books by Michael Argyle

The Scientific Study of Social Behaviour (1957)
Religious Behaviour (1958)
Social Psychology through Experiment (with G. Humphrey) (1962)
Psychology and Social Problems (1964)
The Psychology of Interpersonal Behaviour (1967, 1972, 1978)
Social Interaction (1969)
The Social Psychology of Work (1972)
Social Encounters (1973)
Bodily Communication (1975)
The Social Psychology of Religion (with B. Beit-Hallahmi) (1975)
Gaze and Mutual Gaze (with M. Cook) (1976)
Social Skills and Mental Health (with P. Trower and B. Bryant) (1978)
Person to Person (with P. Trower) (1979)
Social Situations (with A. Furnham and J. A. Graham) (1981)

Social Skills and Work

EDITED BY
MICHAEL ARGYLE

METHUEN
London and New York

First published in 1981 by
Methuen & Co. Ltd
11 New Fetter Lane, London EC4P 4EE

Published in the USA by
Methuen & Co.
in association with Methuen, Inc.
733 Third Avenue, New York, NY 10017

Typeset in Great Britain by
Scarborough Typesetting Services
and printed in Great Britain by
Richard Clay (The Chaucer Press) Ltd, Bungay, Suffolk

British Library Cataloguing in Publication Data

Social skills and work
1. Group relations training
I. Argyle, Michael
305   HM108

ISBN 0-416-73000-0
ISBN 0-416-73010-8 Pbk (University paperback 746)

Contents

Notes on the contributors

Michael Argyle is Reader in Social Psychology at Oxford, and a Fellow of Wolfson College. He has been teaching Social Psychology at Oxford since 1952, during which time he has also held a Fellowship at the Center for Advanced Studies in the Behavioral Sciences (USA) and Visiting Professorships at universities in the USA, Canada, Europe, Israel, Africa and Australia. Since 1963 he has directed a research group at Oxford concerned with various aspects of social interaction, particularly non-verbal communication, social-skills training, and the analysis of social situations. He is a former Social Psychology Editor of the *British Journal of Social and Clinical Psychology*. A list of his publications appears on p. ii.

Michael Brenner is Lecturer in Sociology at the Department of Social Studies, Oxford Polytechnic. He studied at both the University and Institute of Advanced Studies, Vienna, and at the University of Oxford. His research interests are currently in the social psychology of data-collection methods and in the psychology of action. He has edited a number of publications including (with H. Strasser) *Die Gesellschaftliche Konstruktion der Entfremdung*; (with P. Marsh and Marylin Brenner) *The Social Contexts of Method*, *The Structure of Action* and *Social Method and Social Life*; and (with W. Bungard) *Sozialwissenschaftliche Forschung als Soziales Handeln.*.

Nicholas J. Georgiades, who held, until recently, the Headship of the Department of Occupational Psychology, Birkbeck College, University of London, is now Director of the Centre for Leadership and Organization Resources, a non-profit-making organization specializing in training and development activities for both industry and commerce. Over the last twenty years he has been involved with a wide variety of organizations in a consulting capacity, and he was responsible, while at Birkbeck, for the design of training programmes for occupational psychologists. His present clients include British Airways, the National Westminster Bank, Shell International Petroleum in the UK, and the World Bank and associated agencies in Washington, DC.

Christopher K. Knapper is Teaching Resource Person (a general adviser on teaching methods) and Professor of Environmental Studies and of Psychology at the University of Waterloo, Canada. He has written numerous articles and books on various aspects of applied social psychology and the psychology of instruction, including (with P. B. Warr) *The Perception of People and Events*; (with Geis, Pascal and Shore) *If Teaching is Important . . .: The Evaluation of Instruction in Higher Education*; and *The Evaluation of Instructional Technology*.

Robert McHenry is Lecturer in Psychology at Oriel and St Anne's Colleges, Oxford, and Senior Lecturer in Psychology at Oxford Polytechnic. He read Psychology at the Universities of Belfast and Oxford, where he did research on how people form impressions of one another. His present research interests are in techniques of personnel selection and appraisal (including the interview), non-verbal communication, interpersonal attraction and memory for faces. He designs systems of personnel selection and appraisal for industry, and numbers many multinational companies amongst his clients. His books include *Sexual Attraction*.

Ian Morley is Senior Lecturer and Chairman of the Department of Psychology, University of Warwick. His research interests include social psychology, especially cognitive social psychology and the study of group performance, leadership, self-fulfilling prophecies, psychology and the law, social factors in memory, and interviewing. Recently, the SSRC awarded him a grant to study (with A. Sherr) the effects of experience on the initial interview between a lawyer and a client. He has written extensively on bargaining and negotiation, including (with G. M. Stephenson) *The Social Psychology of Bargaining*; he is currently writing a book (with M. King) on *Social Psychology and the Law*.

Vanja Orlans is a Temporary Lecturer in the Department of Occupational Psychology, Birkbeck College, University of London. After several years as a stockbroker in the City of London, she resumed her psychology studies, initially at the Institute of Education and subsequently at Birkbeck College. Her research interests lie in the area of organizational change, decision making and the nature of managerial work. During the last three years she has been involved in the design and implementation of training and development programmes for managers in a wide variety of UK organizations.

S. E. Poppleton is a Principal Lecturer in Psychology at Wolverhampton Polytechnic, and a committee member of the Division of Occupational Psychology of the British Psychological Society. From 1968 to 1970 he worked at the National Institute of Industrial Psychology, where he was primarily concerned with selection and occupational guidance. Since entering academic life, he has been active in both research and consultancy, primarily in the field of selection. He is currently working on the development of a sales-aptitude test. His publications include (with P. Ribeaux) *Psychology and Work*.

Acknowledgements

The publishers and contributors would like to thank the following for their permission to quote tables and figures in the text: *European Journal of Social Psychology* for Table 7.1; Harper & Row, Publishers, Inc. for Figure 5.1; Mouton Publishers for Table 7.2; and Princeton University Press for Figure 4.1.

Introduction

MICHAEL ARGYLE

By 'social skills' we mean the styles of social behaviour used by interviewers, nurses or others in dealing with their clients. As with motor skills — like skiing, typing or driving a car — some people are more skilled than others; they are more effective in attaining the required goals. The effectiveness of different performers must be somehow measured or assessed. Sometimes there are objective indices of success, as in the case of selling; sometimes it is necessary to resort to ratings by supervisors or colleagues. The second step is to compare the styles of social behaviour used by effective and ineffective performers of the skill to discover what they do differently and eventually to define the optimum style of social performance. The adoption of different social skills can have a considerable effect on the attainment of goals. The differences in effectiveness between good and bad performers, or between those at different ends of skills dimensions, are quite often fivefold in terms of measurable goals, e.g. the amount sold by salesmen. At the lower end of the scale, performance can be completely useless: supervisors of groups who produce nothing, psychotherapists whose patients get no better (or get worse) and selection interviewers whose selections are no better than chance. These are all jobs where social performance is of crucial importance; there are plenty of other jobs where it is much less so, such as research and technical positions, though here too it is necessary to be able to communicate and co-operate with other people.

When the optimum social skill has been discovered it can then be taught to trainees on training courses. An implication of the social skills approach is that specific styles of social behaviour will be taught — as opposed to attempts at increasing general sensitivity or insight, as in other approaches to the problem. In the early stages of social skills training and research emphasis was placed on the correct amount of use of elements of behaviour, such as smiling, gaze, head-nods, etc. — and socially inadequate people make less use of these non-verbal (NV) signals, or use them in the wrong way (Trower 1980). Emphasis has been placed on NV signals, since they are important and since trainees are often unaware of the NV signals they are sending, or which are being sent by others. However, verbal behaviour is also highly important, and training can be given for this also.

Awareness of the importance of social skills is fairly recent. Of the skills to be discussed, the first to be studied was probably the supervision of working groups. During the early 1950s, field studies by research workers at the University of Michigan and elsewhere showed some aspects of the most effective style of supervision (Likert 1961). These studies were extensively replicated in many parts of the world and were rapidly incorporated in training courses (Argyle 1980). More details of these skills are given in Chapter 5. At first such courses used the lecture and discussion method, but this was soon found to be ineffective, and was replaced by more powerful training methods (see Chapter 8). The result must surely be that supervisory skills have been changed throughout much of the western world.

The skills of teaching were discovered at a rather later date, but have been if anything even more widely promoted than supervisory skills (Dunkin and Biddle 1974). We have decided to omit them from this book because of the extensive literature on the subject. Some of the other skills discussed have been studied more recently, and some of them need a good deal more investigation.

While knowledge of particular social skills was developing, intensive laboratory research was being conducted into the basic processes of social interaction of which skilled performance consists. Research at Oxford in the early 1960s into social interaction led to the formulation of the social skill model, which draws on the similarities between social behaviour and the performance of motor skills (Argyle and Kendon 1967). Later research on social interaction at Oxford and elsewhere elaborated that model — for example, by showing the importance of NV signals (Argyle 1975). Some of this research is quite recent — e.g. into sequences of interaction and the analysis of social situations — and it has not yet been fully used by those engaged in the study and teaching of specific skills. Research in other areas, such as the analysis of social relationships, is newer still and has not been applied at all. Research in the applied field, however, often makes fundamental contributions to our knowledge of skills. For example, research on sequences of interaction in the classroom (Flanders 1970, and others) has made an important contribution to the study of sequences of behaviour. These processes are described in *Social Skills and Health* (Argyle 1981, Chapter 1).

Since there are a number of different processes used in the production of skilled social performance it follows that expertise requires competent performance in several different areas – accurate perception, effective NV communication, appropriate self-presentation, mastery of skilled sequences of behaviour and so on – and that social performance can fail in various ways; this is found in mental patients, but the same is true of all social skill performers. Although the emphasis of a social skills approach is on skilled *performance*, it does not follow that cognitive factors are being overlooked. Performers must have an adequate understanding not only of the principles of social behaviour but also of the special field of skill in question.

Social Skills and Health (Argyle 1981) includes several topics which may be relevant to managers and others in charge of people at work. For example, it has chapters on social inadequacy and how it can be dealt with, and on the skills needed by social workers and therapists, which are highly relevant to personnel work.

Social skills training (SST) for socially inadequate clients, especially in the USA, has often taken the form of assertiveness training (Rich and Schroeder 1976). However, many socially skilled tasks require forms of influence which have nothing to do with assertiveness. Further, there are cultural differences in the extent to which assertiveness is valued.

This volume is about social skills and work. One of the origins of the social skills approach was the realization that manual workers needed to acquire other skills in addition to manual ones and that some people at work, like supervisors and salesmen, needed to learn social rather than manual skills. It was realized that the concept of social skill could be extended to negotiation, committee work and other managerial activities. Management skills came to be seen primarily as a number of discrete, set-piece performances such as various kinds of interview, each of which could be trained (Sidney, *et al*. 1973). The alternative approach was to give T-group or similar kinds of general-ized sensitivity training in groups which did not include any specific instructions in social skills. The relative merits of the two approaches are discussed in Chapters 5 and 8.

An important problem in using work and other skills is the need to vary social behaviour for different clients and different situations. Research in several areas has shown how skills should vary in this way, and basic research in dyadic interaction has shown how to control the behaviour of others, in interviews or other settings. Recent research into the properties of social situations has shown how skill needs to vary with the social setting. Patterns of social behaviour also vary with class and culture; Chapter 7 discusses the skills of coping effectively in another culture, which is an increasing problem for members of working organizations.

The social skills which are needed at work and elsewhere in society are different at different times and places. The same job may have to be done in different ways as the result of technological or other social changes. The power of industrial supervisors is reduced by the extension of industrial democracy and was much greater before the appearance of trade unions. There may be changes in the law which affect the power and responsibilities of doctors and others. Selection interviewers have a different relation to candidates depending on whether jobs are scarce or good applicants are scarce. There have been general changes in social relationships so that a less hierarchical and less authoritarian style of behaviour is now expected in most organizations.

Social changes also create new social roles, which require new social skills. Technological changes have created the roles associated with television, e.g. anchor men, political interviewers and performers at chat shows; other technological changes have led to the appearance of air hostesses — a role that was deliberately created, and for which training is given. The activities of psychologists led to the roles of psychotherapist, T-group leader and social skills trainer. Now that we know more about social skills, the social skills requirements of new organizational structures could be studied so that new roles can be carefully designed and appropriate training given.

The use of SST has grown very rapidly during recent years — e.g. for mental patients, prisoners, teachers, managers and doctors — though it has not yet become easily available to the general public. Early forms of training, by lecture and discussion, were soon found to be ineffective, and were replaced by role playing, usually with video-tape playback. There have been many follow-up studies of SST, and we can now specify in some detail the form the training should take. These findings are reviewed in Chapter 8. Curiously, those responsible for administering training, in industry and Government for example, have been rather uncritical, have not made demands for follow-up studies before commissioning training schemes, and have sometimes approved unsatisfactory forms of training.

The rapid growth of SST has sometimes led to a low-grade, watered-down form of training consisting of rather amateurish role playing. SST is a sophisticated affair, and will be successful only if full use is made of knowledge of the skills to be taught and of the best techniques of teaching them. Clearly, however, social skills are not the whole story: in addition to the technical knowledge and skills needed, certain kinds of 'personal growth' are needed for those who are going to hold responsible jobs and make difficult decisions.

A number of criticisms have been made of the social skills approach. It is sometimes said that leaders are born and not made, and similar remarks are made of other social skills. Whether there is any *genetic* component of social competence is not known; certainly by the age of 20 some people are very much more socially competent than others, but this is probably from informal, unplanned and chance social learning experiences. However, people can undoubtedly be trained to be more effective performers of social skills, though everyone probably has limits to what they can be trained to do. The same is true of motor skills like performance at sport, though the limiting factors here are chiefly muscular strength and other aspects of physique.

It may be said that a person's effectiveness depends on his power, or other favourable and unfavourable aspects of his situation. These factors are obviously very important, but unfavourable situations can be coped with by using appropriate skills. Fiedler's research (1967) has suggested that a leader whose group does not accept his authority should resort to a different style of supervision. There may be *role conflicts* if conflicting demands are made by other people; for example, a salesman may be under different pressures from customers and colleagues. These require special skills of role bargaining to keep both parties happy. In some roles a number of different people with different points of view must be dealt with; these may be difficult to reconcile, as for social workers who have to deal with the police, doctors, teachers and parents, as well as with their clients.

It is often said that training people in social behaviour encourages deception: pretending to have attitudes and feelings which are not truly felt. Part of the answer is that the rules in some situations and the rules governing a number of professional roles require people to control not only their behaviour but also their emotional states and their attitudes to other people. This can be done by controlling the expression of emotion and by controlling other bodily states, e.g. by relaxation, and thoughts and images (Hochschild 1979). Even if such control of feelings and attitudes is unsuccessful it can be argued that teachers, doctors and others should still treat well, i.e. give the appearance of liking those clients whom they do not like. There is some evidence that real feelings change to fit those which are expressed (Laird 1974).

Another objection to the social skills approach is that people may be made self-conscious by being instructed in the details of social performance. The experience of trainers is that this is only a temporary phenomenon; after the second training session most trainees focus their attention once again on the job in hand and the behaviour of the others present rather than on their own perform-ance. Finally it is sometimes objected that the use of skilled social techniques is a form of 'manipulation' of others. This is a curious point: if a teacher teaches effectively, or a doctor cures patients, this would not be regarded as manipulation, which presumably refers to successful social influence of a kind which is thought socially un-desirable. Perhaps the use of subtle, or non-verbal, social skills is regarded with more suspicion than the use of more obvious, or verbal, skills. The social skills approach extends the range of social techniques beyond those which are familiar. It is hoped that the new skills will be used more for desirable social ends than for undesirable ones.

References

Argyle, M. (1975). *Bodily Communication*. London: Methuen.
_____ (1980). The development of applied social psychology. *In* Gilmour, R. and Duck, S. (eds). *The Development of Social Psychology*. London: Academic Press, pp. 81–105.
Argyle, M. (ed.) (1981). *Social Skills and Health*. London: Methuen.
Argyle, M. and Kendon, A. (1967). The experimental analysis of social performance. *In* Berkowitz, L. (ed.). *Advances in Experimental Social Psychology*. New York: Academic Press, vol. 3, pp. 55–98.

Dunkin, M. J. and Biddle, B. J. (1974). *The Study of Teaching*. New York: Holt, Rinehart and Winston.

Fiedler, F. E. (1967). *A Theory of Leadership Effectiveness*. New York: McGraw-Hill.

Flanders, N. A. (1970). *Analyzing Teaching Behavior*. Reading, Mass.: Addison-Wesley.

Hochschild, A. R. (1979). Emotion work, feeling rules and social structures. *Am. J. Sociol* 85, 551–75.

Laird, J. D. (1974). Self-attribution of emotion: the effects of expressive behavior on the quality of emotional experience. *J. Pers. Soc. Psychol.* 29, 475–86.

Likert, R. (1961). *New Patterns of Management*. New York: McGraw-Hill.

Rich, A. P. and Schroeder, H. E. (1976). Research issues in assertiveness training. *Psychol. Bull.* 83, 1081–96.

Sidney, F., Brown, M. and Argyle, M. (1973). *Skills with People: A Guide for Managers*. London: Hutchinson.

Trower, P. (1980). Situational analysis of the components and processes of behavior of socially skilled and unskilled patients. *J. Consult. Clin. Psychol.* 48, 327–39.

1 The selection interview

ROBERT McHENRY

Introduction

There is a tendency to see selection interviewing as best practised by someone who is a good judge of people. If that person tends also to be a good conversationalist, so much the better. I share neither of these points of view. Good selection interviewing depends mostly on the efficient use of a system for selection in which the interview itself plays only a small part. Equally, the social skills of the conversationalist mean little if the interviewer practising them has almost no idea about what he is looking for. The chapter will attempt to put the interview where it belongs: a thin slice between the careful preparation that must precede it and the careful consideration that must follow it. What follows, therefore, is a description of a system rather than a conversation by itself.

Job selection is an extremely difficult task even for experts who spend large amounts of time and gain large amounts of information about potential employees. Carefully planned selection programmes such as the choice of American Peace Corps volunteers are not without their failures (see Wiggins 1973, pp. 580 ff.). the reason for the difficulty is not hard to see. In effect, what selectors are trying to do is to predict the behaviour of each person in the range of situations that go to make up a particular occupation. Many methods have been used in attempting to predict individual behaviour. Among the most notorious are personality tests. Mischel (1968) has argued that all of the evidence indicates that personality tests fail dismally in this application. The up-to-date view about how selectors should predict the behaviour of candidates is based on the notion that *the best*

*predictor of a person's behaviour is his behaviour in a similar situation
in the past*. This is far from being an infallible rule but it is much
more likely to be a better guide to predicting behaviour than any
other assumption. A good selection system should therefore begin
with an examination of the job to be filled and from this is derived a
list of the most important situations a job holder is expected to
perform well in. Candidates are interviewed to discuss how they have
behaved in situations similar to these in the past. Information is then
collated in a way to be described later and candidates compared on
the basis of their past performance. This recommended selection
procedure will now be described as a three-part sequence.

Preparation for the interview

To decide on the important situations which need to be filled on any
particular job the selector needs to look at a *job description*. Many
organizations provide job analyses and descriptions as part of their
personnel services and the fortunate selector may thus find his job
made easier. Those less fortunate will have to make their own
descriptions of the vacant job. This is usually not a difficult task. In
essence, it is a comprehensive listing of the duties of the job together
with a note of their importance for the organization and of the job
time they are expected to occupy. Such an exercise is necessary
because it is easy to forget the whole range of job duties and to get
some of them out of proportion. For example, members of my profession do not simply lecture; they also perform a wide range of management, research and administrative roles. The job description can be
done on a purpose-designed form and Table 1.1 shows such a form.

The form used in Table 1.1 has been completed with details of
a relatively simple job. The vacancy is for a general clerical
assistantship in a small office. The assistant is expected to do the filing
as well as perform telex, post and reception services. The remainder
of the form is a listing of the main duties under three separate
headings. The layout of the form is as simple and unfussy as possible
in the hope that it will encourage nearly all selectors to use it.

Job description leads to *job-holder specification*. The goal of this
exercise is to use the comprehensive job description to decide what
kind of candidate would best fill the job. Unfortunately, this is not
simply a matter of listing the 'qualities' of the acceptable candidate. It
is now widely accepted that the idea of 'qualities' or 'traits' in relation

Table 1.1 Job description

JOB TITLE

General
Clerical Assistant

TITLE OF MANAGER

Office Supervisor

MAIN AIMS OF JOB (not more than 40 words)

Provide filing, telex, post and reception service

WORKING CONDITIONS (e.g. travel, unusual features)

Continuous office environment

RELATIONSHIPS TO PEOPLE (e.g. inside and outside company? Individuals or groups? Type of relationship?)

Receives visitors
Answers telephone when salesmen are busy
Acts as general assistant to manager
'General assistant' relationship to staff

RESPONSIBILITIES AND DECISIONS (e.g. for negotiations, money, decisions about people, commitment of resources, forward estimates, care of plant, buildings, equipment, quality, etc.)

Operate telex machine
Distribute telex messages
Distribute mail
Make outgoing mail ready
File correspondence and control documents
Make refreshments
Receive visitors
Take telephone messages

---

to individual behaviour is a misleading one (Mischel 1968). To say of an individual that he is a 'leader' or an 'extravert' is common in everyday descriptions of individuals but has no place in profession selection. Behaviour is not general but it is tied closely to social situations. Therefore, a 'leader' may lead in one situation but not in another; an 'extravert' may show extraverted behaviour in one situation and introverted behaviour in another. It is very difficult to predict individual behaviour from one situation to a different one and for selectors to do so unthinkingly is sloppy and dangerous. The solution is to make out a job-holder specification, not just in terms of the behaviours required but also the situations in which these behaviours are expected to be displayed. 'Behaviour' is a key word here. Selectors should be interested in specific past behaviour or action in specific situations. He must discover where and when a candidate has referred to behaviour in any particular way. The same principle applies to all behaviour to be predicted. The interview and subsequent assessment will depend on the precision and care taken at this preparatory stage.

Job-holder specifications may be made on specially prepared forms (see Table 1.2). In the example illustrated, behaviour is listed under five main headings. These are:

(1) General qualifications.
(2) Skills.
(3) Social roles.
(4) Pressures.
(5) Interests.

The example is based on the general clerical assistantship vacancy which was outlined earlier. This time the information is organized according to the tasks expected of the job holder and the settings in which these tasks are to be performed. Obviously, even more detail could be included but it is probably not necessary because the interviewer himself will know which office equipment is used and which 'paper-work systems' are in operation, etc. What has quite properly been avoided is the general and almost useless description of the desired job holder in terms like 'intelligent', 'reliable', 'hardworking', 'sociable' and 'stable'.

The occupant of a job is often expected to perform in several different situations, e.g. teaching, research and administration in the case of a university lecturer. In such cases different skills or abilities

Table 1.2 Job-holder specification

JOB TITLE	
General Clerical Assistant	

NAME AND TITLE OF MANAGER	
Office Supervisor	

GENERAL QUALIFICATIONS	Lives within easy travelling distance. CSE-standard English and Commerce or clerical experience with good references. Reliable in attendance and time-keeping. Clean and tidy.
SKILLS	Methodical and accurate at filing. Pleasant speaking manner. Dexterity on office equipment (telex, works with several paper-work systems).
SOCIAL ROLES	Deal with a variety of visitors (mostly men). Speak to a variety of people on telephone (usually customers). Mix with a small work group (mainly aged 18–24). Accept assistant role with sales staff and manager.
PRESSURES	Ability to switch jobs quickly when requested. Ability to tolerate small, enclosed work groups. Acceptance of routine with only a little variety. Ability to cope with occasional rush jobs.
INTERESTS	Likes office work. Likes stability and relatively invariable work. Possibly seeking advancement. Ability to tolerate intellectually undemanding work.

are needed and it may be necessary to decide which are most important, and whether or not a minimum competence is necessary in each area.

Once the interviewer has got as far as preparing a job description and job-holder specification, he has done most of his preparation. However, there are other aspects of preparation such as recruitment advertising and writing letters to applicants, which will not be dealt with here. After these, the selector's next step is to preselect candidates for interview. The most helpful documents here are likely to be the job-holder specification and a set of completed application forms. Unfortunately, many otherwise satisfactory selection systems fail at this point because a poorly designed application form does not give the selector enough relevant information on which to base a judgement of whether to interview or not. Indeed, in my experience of selection systems the application form is often the weakest link in the chain of selection documents. The very least an application form should do is to ask applicants to describe their present job and to list from this and any other experiences the activities relevant to the job they are applying for. If this is done applications from dissimilar-sounding present jobs and those who appear initially to have no relevant experience may not be so easily passed over. The object of preselection is to avoid wasting interviewer time and to ensure that only the most likely candidates are seen. A good rule of thumb is to preselect no more than four candidates per vacancy. This is based on the observation that good interviewing is time consuming and demands concentration. It is difficult to carry out selection efficiently if the number of candidates means that time with each one must either be kept to a minimum or that successive interviews will span several working days.

The next step is for the selector to decide which of his colleagues should help with the interviews. He will need the services of at least one other person and probably a maximum number of three for reasons which will now be made clear. The best interviews are conducted 'one-to-one', i.e. the candidate and only one interviewer present during the interview. The single interviewer establishes best rapport and so can elicit a more subtle and frank exchange of information. Many organizations (particularly universities and local government) prefer (quite wrongly) to use an interviewing panel, all of whom are present in the same room for each interview. Few interviewees find that establishing rapport with several people across a table is easy and such panels run the risk of not allowing the candidate to give a valid impression. Even if the candidate feels on terms with all panel members it is nevertheless a nerve-racking experience to impress several people at the same time. The panel interview, more than any other situation, favours the confident candidate able to think and talk convincingly. It is missing the point to say that these are desirable qualities in any candidate: all jobs use more than just 'acting' ability. On the other hand, panels are believed to be less biased because more than one person judges the candidate. Why not try to marry this advantage of panels to the advantages of a one-to-one interview by having two or three interviewers see each candidate in turn? They can all judge independently as before and compare notes after all the interviews have been finished. This need not take up any more time on the part of the interviewers, for they can all be seeing different candidates during the same time and changing over at the end of a prearranged interval. It might be objected that it is hard on the candidates to be interviewed three times for the same job. However, in my experience candidates are only too well aware of the caprice of panel decisions and welcome having more than one chance to discuss their abilities, especially if it is within the intimacy of the one-to-one interview.

There is still more to do before the interviews themselves. One of the more important tasks is to plan the interviewing day so that there is at least 15 minutes of free time between interviews. This will be needed to make notes on the previous interview and to refresh the memory about the next. Also, the applications of preselected candidates need to be re-read and the more important points transferred to a small card or sheet of paper in brief note form. Whilst doing this, the interviewer should plan his sequence of interview topics. There are two good general plans. If the candidate is young and has a short job history it is best to work chronologically *forward* during the interview, starting at his or her school career and working up to the job held at present. Otherwise, it is preferable to work chronologically *backwards*. Older candidates often fail to see the relevance of anything but their present job to the job to be filled and for this and other reasons it is a good idea to start at this point.

The final and smaller preparations need to be done just before the interview itself. As it is bad enough to suffer one interruption during an interview, let alone several, a large notice should be placed on the door of the interview room asking that the occupants be left undisturbed.

The same must apply to the telephone, even if it means placing it in a drawer, leaving it off the hook or just muffling it with a coat! Finally, some attention must be paid to the layout of the room. Most interviewers use their own office and there is nothing wrong with this, providing they do not sit behind their desk. Room layouts are critical to the rules of the conversations that take place within them. By far the best layout is for two equal-status chairs to be placed at right angles about 1.5 m (3–4 ft) apart. This defines an 'easy' atmosphere, which means one in which the candidate will be encouraged to talk to the ultimate gain of the selection process.

The interview period

Most people have definite expectations about what should happen in the selection interview and may be very disturbed if these expectations are not met. Contemporary conventions are:

(1) Each interview would consist mainly of the interviewer asking the candidate questions about previous experience. These questions may be awkward or personal, as long as they are relevant to the job.
(2) All parties will treat the interview as a serious and formal exercise and the interviewers will dress and act in the manner they expect of candidates.
(3) At the end of the interview candidates will be given a precise idea of when to expect a decision.

The discussion of preparation for the interview will have given the reader some idea of what the interviewer is expected to look for. The interviewer's task is to encourage the candidate to talk about his experience and expectations and to seek, in the information obtained, evidence of recurring patterns of behaviour in particular situations. As mentioned earlier, the best predictions of future performance derive from perceptive recognition of persisting elements in the candidate's past behaviour in situations similar to those in which he or she will be placed should they get the job. The interview should be thought of as passing through four stages. These are:

(1) *Rapport establishment* – During which introductions are made and the interviewing procedures are explained.

(2) *Biographical questioning* − In which the interviewer takes the candidate through his or her biographical record.
(3) *Answering queries* − In which the interviewer gives the candidate any facts he or she needs.
(4) *Parting* − In which both sides agree on subsequent procedures.

These four stages will now be dealt with in turn.

Rapport establishment

Many interviewers take rapport to mean 'chumminess', and so begin an interview with a long discussion on, for example, the weather, Saturday's football and the candidate's hobbies. Rapport is not chumminess; it is the establishment of a business-like relationship. The best thing an interviewer can do is to tell the candidate briefly how long the interview is likely to last, that he will do most of the questioning and that there will be plenty of time left at the end of the interview should the interviewee have anything he would like to raise. All this can be achieved in a couple of sentences. Some interviewers try to discuss common acquaintances, background or interests to establish a basic relationship with the candidate. The most important thing is to get the interviewee to start talking and get him going on something relevant and interesting. Therefore, it is a good idea to ask him to discuss immediately the most satisfying aspects of his present job.

Biographical questioning

During the interview, selectors will make use of the four basic 'tools-of-the-trade'. The most important of these is the *open-ended question*; questions worded so that they suggest lengthy replies. Such questions usually begin with 'How . . .?', or 'Why . . .?', etc.: 'Why did you leave that job when you appear to have been doing so well?' or 'How have you tackled that particular problem in the past?'

One difficulty most interviewers have is that the formulation of such questions does not come easily to them. In everyday life we all tend to use another form of question called the *closed-ended question*. The definition of such a question is that strictly speaking it demands a short reply such as 'yes' or 'no'. In everyday conversations we hardly bother about this, because of the convention that 'we make conversation', and so when we are asked a closed-ended question we usually embark on a fairly lengthy reply. This everyday rule is not always carried over to a selection interview. As a result, an interviewing sequence can often go like this:

Interviewer: 'I see you went to school in Cardiff.'
Candidate: 'Yes.'
Interviewer: 'Was it pleasant there?'
Candidate: 'Yes.'
Interviewer: 'And after you left you went to work at Evans and Jones?'
Candidate: 'Yes.'
And so on.

The solution to all this is for interviewers to practise using only open-ended questions. There are plenty of opportunities for this in everyday conversation. When watching television, another idea is to reword the questions asked by professional presenters (most of whose basic interviewing skills are very poor indeed). Another basic element in the interviewer's repertoire is the *reflection/summary*. A good interviewer demonstrates his interest and attention to the candidate by occasionally reflecting back the impression that his statements are giving. Thus:

Interviewer: 'So you see yourself as the kind of man who likes to work under pressure?'
Candidate: 'Yes, most of the time.'

Reflections have an additional advantage in allowing a candidate to correct an interviewer who has the wrong impression. *Summaries* are reflections that come at the end of a 'paragraph' in the interview and they are used to complete one section neatly before moving on to the next. For example:

Interviewer: 'And in the final months you spent in that job, you were feeling more and more that you would like more responsibility.'
Candidate: 'Yes, that is true.'
Interviewer: 'Well, how did you find it in the job you moved on to next, were you given more responsibility?'

A summary here allows the candidate to see that one part of his experience has been finished with and that the interviewer is moving on to the next.

Another essential element of the interview is the *pause*. Many experienced interviewers are often more embarrassed by silence than candidates are. However, it is important to learn that having asked a question the interviewer must be able to wait a reasonable amount of time (say, up to 45 seconds) until it is answered. If the interviewer is deliberately asking a difficult question it is a good idea to prefix it with a warning such as 'Look, this is a difficult question, so please take your time'. This will relieve some of the pressure that silence exerts on candidates.

The interview is a blend of the above techniques but the most important of them is the open-ended question. Open-ended questions, especially at the beginning of the interview, encourage candidates to talk. The closed-ended question should be used very sparingly and only to check on facts. It is difficult in this written account to convey a clear idea of the conduct of the good biographical interview. To the casual listener it would sound rather like a very informal chat. Good interviewers need to listen carefully to what the interviewee is saying and where possible to link the next question to the statements or points the interviewee has just made. This keeps momentum going and the practised interviewer will soon find that he can get interviewees to talk for almost 80 per cent of the interview time. An interview which achieves a different balance from this is probably not going as well as it might. The questions an interviewer asks should all be related to finding out about the interviewee's experiences which are similar to those written down on the job-holder specification. Some interviewees will speak spontaneously about these, particularly if the selector has given them a job description to read before the interviews take place. With other interviewees it can be more difficult to get into relevant areas. However, by asking questions on present and previous jobs the interviewer will soon find openings which he can explore in depth to find out how the interviewee has behaved, how much help he has required with certain tasks, what difficulties he commonly gets into, etc. There are no prescribed questions. It is up to the interviewer to probe the interviewee to discover facts about his behaviour and the circumstances under which it occurs. He might start with 'Tell me what you do on a typical day?' and move on to closer scrutiny of the likes and dislikes of the candidate while at work. Something the candidate may say will suggest a hypothesis to the interviewer (e.g. the candidate dislikes taking orders from younger people). This can then be explored by finding if the candidate's reaction is the same in similar situations.

There is one note of caution to be added: interviewers should never be tempted to ask hypothetical questions. It is wrong to pose questions such as 'Suppose the photocopier were to break down and could not be mended for a week; what would you do?' Faced with questions like these interviewees may simply give the answer they think the interviewer wishes to hear — or they may even state honestly their *intentions*. However, the gap between intention and actual behaviour may be very wide. That is why past behaviour makes a much better basis for a behavioural prediction.

One of the fundamental problems of interviewing is that it depends mainly on verbal reports from candidates, which may be inaccurate because they want to create a favourable impression and because they do not have accurate insight into or memory of past events. Self-presentation biases can be overcome to some extent by persistent and detailed follow-up questions of the kind described. The other kind of error is more difficult to deal with. There are, however, some familiar sources of error here: for example, people tend to attribute their successes to themselves and especially to their own efforts, whereas failures tend to be attributed to other people or to bad luck (Schneider, *et al.* 1979).

With all this information about past behaviour to be remembered, almost every interviewer will wish to take notes. There is nothing wrong with this as long as they are only one- or two-word reminders. The best time to make full notes is immediately after the interview, which is why a minimum gap of 15 minutes between the finish of one interview and the start of the next was recommended earlier. Further reference to note taking will be made in the next section.

Non-verbal skills in the selection interview

Verbal skills, such as forming and asking the right questions, have already been dealt with. A less important aspect of the selection interview is the appearance and manner of the interviewer. These 'non-verbal' (NV) skills are relegated to a secondary role simply because most interviewers possess the requisite skills to a high degree. However the interviewers must be reminded to use these skills in the interview itself and to understand how NV behaviours can control the flow of a conversation and the rapport between the interactors.

NV behaviours in the interview may be conveniently divided into three classes: signals, cues and regulators.

Signals are NV signs made deliberately by the interviewer to define a situation and to maintain rapport with the interviewee. The most obvious of these is the layout of the interview room itself. Rooms (particularly offices) and the arrangement of furniture within them have been the subject of much comment and research (Joiner 1969, Korda 1976). Both of these authors suggest that most offices may be divided up into two different areas. Korda calls them 'pressure' and 'semisocial'. The pressure area is the zone around the desk. It is a setting for a formal conversation in which the usual occupant of the office sits behind the desk and takes the lead. The semi-social area is an area away from the desk in which there are armchairs and a coffee table. Here, the conversation takes place on a more equal basis between the interactors. According to Korda, the astute businessman will choose to use that area of his office most appropriate to the encounter he expects to have with his visitor. There is a hint here for the interviewer who wishes to carry out an efficient biographical search. The best setting for such a search is obviously the less formal 'semi-social' area, for it is in these surroundings that an interviewee will be best encouraged to talk freely. Within this area the best position seems to be for the interviewer and interviewee to sit at right angles to each other. Sommer (1959) found from observing conversations in a systematic way that such a 'corner' position produced six times as many spontaneous conversations as face-to-face positions and twice as many as a side-by-side arrangement. Additional advice may be derived from the observations of Hall (1966) which suggest that a distance of about 1.5 m between interviewer and interviewee is optimum. In Hall's terms, this, for our culture at least, is just about the boundary line between what can be regarded as the 'personal distance—far phase' and 'social distance—close phase'. In other words, this distance will be perceived by most as being neither too friendly nor too formal.

Another NV signal of great importance to the interviewer is his use of gaze. In everyday conversations we look at someone to let them know we are interested in what they are saying. In my experience, such a response comes easily to all but the nervous or badly prepared interviewer. Such interviewers can spend a lot of their time gazing at their notes or at an application form. The interviewees find it un-rewarding to talk to someone who does not appear interested in what they are saying. Moreover, the interviewer misses *cues* from a candidate's gestures and face and fails to notice NV *regulators* (the importance of which is discussed below). The moral is to prepare well for interviews so that the necessity to look at papers during the interview itself may be avoided. In fact, it is best to do away with any distracting papers and bring to the interview itself only the briefest of notes.

A third kind of signal is NV encouragement, such as smiles, head-nods and 'mmhmms'. A good many interviewers observed by myself have behaved in an unnecessarily formal and serious way during the interview: they adopt stiff bodily postures and inscrutable facial expressions and are rarely aware of this unless given the opportunity to see for themselves on videotape afterwards. This is unfortunate at any time but especially so with candidates who are excessively nervous. It is important to use the ordinary conventions of skilled conversation and give the usual amount of NV encouragement in the selection interview. There is little empirical evidence on the exact effects of reinforcement in conversation, although one study by Siegman (1976) has suggested that 'mmhmms' made by interviewers do not lead to more talking by interviewees. The same study did show that interviewees were always aware that such reinforcers were being used. Another example of an NV reinforcer is gaze. It is a well-recognized cultural rule that the listener looks at the speaker most of the time that he is talking. Listeners who do not look are regarded as rude and uninterested. Finally, the taking of notes during an interview can, if skilfully done, become an NV reinforcer. When interviewees are discussing their major achievements interviewers may take the opportunity to say 'Very interesting. I'll just make a note of that'.

The second general area to be reviewed here is that of NV cues. Cues are (largely unintentional) behaviours of the other person which can often give the watcher an idea of that person's motives, feelings or intentions. The face is the richest source of cues to someone's feelings, for it is controlled by a great number of small muscles and is consequently capable of a wide repertoire of expressions. There is also some evidence to suggest that the facial expressions associated with common emotions (fear, anger, happiness, etc.) are universal in humans and are innate (Darwin 1872, Ekman and Friesen 1969). Ekman and Friesen (ibid.) suggest that if someone is feeling an emotion the clue to that emotion will appear in the face. However, the watcher may have to be vigilant if he wishes to see it for most people learn to mask, neutralize or deintensify their naturally occurring facial expressions. The attempt to do this will often not be totally successful and the result will be a *blend* of the unwanted emotional expressions and the masking or neutral expression. Thus the eyes may retain the look of anger although the mouth is forced into a smile. According to Ekman and Friesen many people will be fairly successful in disguising emotion in the face while being interviewed. However, a second theory of theirs states that emotional expression nevertheless needs to be relaxed and that its 'energy' moves down the body to the outlet nearest to the face – which happens to be the arms and hands. Thus, hand movements and gestures will often be a clue to someone's feelings. Finally, because many people are also aware that their hands are giving them away they learn, in times of stress, to control these also. However, the emotional expression now emerges in the legs and feet and the fact that someone is feeling an emotion may often be detected from movements of that part of their body.

Ekman and Friesen's theory has some application for the selection interviewer because it suggests that signs of tension will appear in the lower parts of the body. Unfortunately, it is not easy, in individual cases, to identify the reason for the tension or the emotion being expressed, simply because the hands and feet are far less specific in their movements than the face. However, the experienced interviewer may be able to make a guess (from excessive lower-body movements) about the nature of an interviewee's feelings on a certain subject and these may be later cross-checked during the same interview.

Cues about an interviewee's present state in the interview (tense, relaxed) may be given by his posture. There is some evidence (Mehrabian 1968) to suggest that formal postures (interviewer sitting up straight with arms folded) or informal ones (sitting back in chair, arms by the side) may respectively indicate tense and relaxed moods. Also, sitting upright with folded arms may make it difficult for someone to relax, because feedback from their own bodily posture suggests to the brain that they are not relaxed and thus a vicious circle is set up. If this is true it is important for interviewers themselves to adopt an open and relaxed posture in the hope that interviewees will imitate them. The corollary is that they should be vigilant in case it is they who imitate the *interviewee's* tense posture! It has been found that when interviewers give NV signs of approval, candidates become more friendly and relaxed and create a good impression with the interviewers (Keenan 1976).

Candidates like interviewers who adopt a more positive, friendly style, which affects whether they will accept offers of jobs or further interviews. The proportions who do accept vary very widely between interviewers. The impression created by an interviewer can also be good public relations for the organization, since candidates will tell their friends what happened.

Several studies have shown that if candidates look, smile and nod their head more they will be evaluated more highly by interviewers (e.g. Forbes and Jackson 1980). However, Hollandsworth, *et al.* (1979) found that appropriateness of speech and the fluency and clarity of speech had more effect on interviewers' evaluation than visual NV signals.

Regulators are the third important category of NV signals. They derive their name from the functions they perform in controlling the flow of conversation. The regulators of most interest to the interviewer are the 'turn-taking signals' used by both participants at points when one speaker stops talking and the other starts. According to one investigator (Kendon 1967) gaze plays an important role here. Kendon's observations of face-to-face conversations showed that, during most of the time that someone was talking, they did not look at the listener but looked away and glanced at the listener only occasionally. However, just as the speaker was about to finish what he was saying he would look the listener in the eye for a much longer period. This acted to the listener as a signal that it was his turn to speak. Kendon made his observations on samples of undergraduates and his account of their use of gaze has stood for many years. However, it has recently been challenged by Beattie (1978) and by Rutter, *et al.* (1978) who found that the long terminal gaze often did not occur at smooth switches, and made a difference only during periods of hesitant speech with low gaze. One or more of these groups of investigators may have observed unusual samples but, at any rate, there is now a doubt that terminal gaze is a universal and important regulating signal.

Some recent investigations by Duncan, *et al.* (1979) suggest that at least one (and probably more) of a list of six NV behaviours are used to indicate to the listener that it is his turn to speak. The six are:

(1) Deviation in intonation of the voice from its level pattern.
(2) An utterance such as 'you know'.
(3) The completion of a grammatical clause using a subject—predicate combination.
(4) A drawl on the final syllable.

(5) Termination of a hand gesticulation or the relaxation of a tense hand position such as a fist.

(6) A decrease in loudness or pitch.

Duncan observed pairs of North-American speakers and it remains to be seen just how many of these regulators are used in western Europe.

The point of discussing regulators and other NV signals and cues is not to teach interviewers how to use them. All interviewers will be used to emitting and detecting these signals (almost certainly without awareness) in everyday conversation. However, what the interviewer has to guard against is getting himself into a position where he fails to detect these signals when they are given by the interviewee. An obvious case occurs when the badly prepared interviewer tries to catch up on his preparation during the interview itself by reading notes or an application form. His failure to look sufficiently at the interviewee has many important consequences. One of them is that he fails to pick up some of the regulating signals and there is consequently an unfilled pause every time the interviewee finishes speaking. The conversation then fails to 'mesh' (Argyle and Kendon 1967) and can therefore become an unrewarding and frustrating experience — especially for the interviewee. Another reason for describing NV behaviour in the interview is that it can lead to a better understanding of how to deal with 'difficult' candidates (e.g. those who talk too much or too little).

This discussion of NV behaviour in the interview has not touched on the idea that an interviewee's postures and gestures can themselves give clues to his or her 'personality'. In the view of the writer, not only are such ideas implausible but they are untested by any method that would withstand close scrutiny.

The interviewer's NV behaviour provides part of the key to another set of interview problems: dealing with 'difficult candidates'. There are several familiar types of difficult candidate, and social skills research can suggest how to deal with them. Examples are:

(1) Candidate talks too little: give NV reinforcement, ask open-ended questions, do not talk too much.

(2) Candidate talks too much: withhold NV reinforcement, ask closed questions, interrupt.

(3) Boastful, bombastic: ask persistent, detailed questions.

(4) Very nervous: extend period of rapport establishment; adopt very relaxed, friendly manner.

Interview bias

Rather a lot has been written in academic papers on bias in the selection interview. It has always been a favourite topic for research and there are far too many references to be dealt with in the space available here. Perhaps this is just as well for, in my view, this type of research is probably not very important for two main reasons. These are that much of the research has been performed in contrived situations; and that researchers' preoccupation with interviewer fallibility is all part of the tendency to give interviewers *negative* advice.

Much of the research has been carried out in contrived situations. It is very unusual in published experiments to find that interviewer and interviewee meet face-to-face; instead, non-expert interviewers see films or read written scripts of the answers given by 'candidates in selection interviews' who are actors rather than real applicants trying to secure a job. This objection to 'interviews' of paper people has been made before, notably by Gorman, *et al.* (1978). Their study enabled a direct comparison to be made between 'paper' and 'real' interviews and there is nothing tentative about their conclusion:

'The two studies together suggest to us that there are profound differences between "interviews" of paper people and interviews of real people. For our part, we consider these differences to be so profound and so pervasive that it seems unlikely that the scientific community will learn anything about the process of interviewing real people from the paper-people analogy' (p. 191).

However, there has been quite a lot of research into sources of bias in real interviews, a tradition started by Webster (1964). Some of the main sources of error are as follows:

(1) Interviewers come to like or dislike a candidate and this affects their evaluation of his suitability for the job. Keenan (1977) found liking and evaluation of suitability correlated at $r = 0.51$, and if the candidate was going to work with the interviewer $r = 0.75$. This correlation was somewhat less for trained interviewers (Keenan 1978).

(2) Interviewers like, and evaluate as suitable for the job, candidates who are similar to themselves in attitudes or in educational or social background.

(3) Interviewers attach too much importance to a candidate's performance in the interview – which is normally quite unlike the job. They attribute his behaviour too much to his personality, not enough to the situation: 'the fundamental attribution error' (Jones and Nisbett 1972).

(4) Interviewers look for, and place too much emphasis on, negative information about candidates and do not attach corresponding importance to positive information.

(5) Interviewers are affected by stereotypes about the abilities or other characteristics of candidates from certain social, racial, regional or educational backgrounds. Such stereotypes are often incorrect and in any case imply that all members of a category are more similar than they really are.

(6) Interviewers are also affected by physical cues such as physical attractiveness, beards, spectacles, height, clothes, etc. which have rather little correlation with the abilities being sought.

(7) Interviewers are influenced more by information which becomes available early in the interview, and often decide about the candidate in the first few minutes.

(8) Interviewers are affected by contrast effects: a middle of the range candidate seen after two very weak ones is evaluated unduly highly.

These and other errors are discussed further by Cook (1979), Schneider, *et al.* (1979) and Schmitt (1976).

Researchers' preoccupation with interviewer fallibility is all part of the tendency in this area to give interviewers *negative* advice. McHenry (1972) has argued that it is impossible to eliminate bias in the judgements one person makes of another. They can, however, be minimized by teaching an interviewer exactly what to look for in a candidate. There is evidence that most of these errors can be reduced by training.

To minimize the more obvious effects of bias on interviewer judgements, two fairly simple steps should be carried out. These will encourage interviewers to collect *evidence* systematically to support their judgements. First of all the interviewer should take time to fill in a form about the candidate. An example of such a form is shown in Table 1.3. There are two main parts to this form. On the right-hand side there is space for interviewers to note the numerical rating they give to the candidate in each of the five areas mentioned before.

Table 1.3 Interview record form

CANDIDATE: VACANCY:	
	RATING
GENERAL QUALIFICATIONS	
SKILLS	
SOCIAL ROLES	
PRESSURES	
INTERESTS	
INTERVIEWER: DATE:	

On the left-hand side are five boxes which are much more difficult to complete; it is here that interviewers are expected to record the evidence gained from the interview or application form which enables the numerical rating to be made. The interview record form is complementary to the job-holder specification form shown in Table 1.2. The selector ought to place the two forms side by side in front of him, job-holder specification furthest from his writing hand, and when writing out his report make constant reference to what he was supposed to be looking for.

Within any area there may be conflicting evidence to be reconciled. Indeed one of the purposes of the interview is to deal with such conflicts of evidence. It is then necessary to combine the ratings in different areas into a single evaluation of suitability for the job. Jackson, *et al.* (1980) found that untrained students were quite good at weighting characteristics appropriately for different jobs — they realized that different qualities are needed by accountants and foremen, for example. Interviewers must be carefully briefed on the correct weights for each ability or trait for the job.

A second step which might be taken to reduce bias is to hold a selectors' conference. An ideal selection system uses two or three trained interviewers working independently of one another and seeing candidates 'one-to-one'. The selectors' conference is the stage at which they combine their judgements. Before the conference starts it is assumed that all selectors have completed an interviw record form for each candidate. After that, the conference is run according to the following rules:

(1) A chairman should be appointed.
(2) At the beginning of the conference, the chairman should call on each member of the conference to give his rating of the candidate in each area of behaviour.
(3) The chairman should record all ratings from all selectors on a master sheet, without comment and without asking for explanations. This procedure ensures that selectors are committed to a point of view which they will expect to justify. It also ensures that the areas of dissent and agreement are made clear from the outset.
(4) The chairman then conducts discussions about each candidate, concentrating on areas of disagreement. He asks in particular:
  (a) On what evidence the ratings are based. This may indicate

that the selectors have different evidence and when their
evidence is pooled they may be able to agree on a unanimous
rating.

(b) Alternatively, selectors may be found to have the same
evidence and to be interpreting it differently. Here the
discussion concentrates on what is a reasonable and generally
acceptable interpretation of the information.

(5) The chairman records the unanimous views of the selectors or a
majority view. He must also decide whether to record any single
strong dissenting view, or any dissent which has a substantial
minority support.

(6) The chairman summarizes for the selectors the main points of
their argument for and against the candidate and these should be
agreed as the basis for the report on the candidate.

By using two or more independent ratings of each candidate and by
combining them in this way, bias for and against certain candidates
will hopefully be minimized. Of course, many selectors' conferences
reach unrepresentative decisions about candidates or reach decisions
which constitute no more than an uneasy compromise between the
selectors. Discussion is often unduly influenced by the relative
strength of the selectors (in terms of seniority, self-confidence, ability
to express himself, freedom from fatigue, preoccupation, etc.).
Sometimes political bargaining may also be used. These are dubious
bases on which to appoint or reject a candidate; they may result in
injustice to an individual and be detrimental to the employing
organization.

Validity of this selection scheme

Is it worth while going to the trouble of applying the system that has
been outlined? Does it work; and if so where is the proof? These
questions must be initially answered rather defensively. It is very
difficult to test the effectiveness of a selection system. For one thing, it
would need to be tried out for its usefulness in selecting people for a
wide range of jobs, which in turn would need an initial sample of
several hundred applicants. Secondly, rival systems (e.g. use of panel
interviewers, personality tests, etc.) would have to be used in parallel
to discover later which gave the best predictions; it would therefore be
necessary for applicants and selectors to give a lot of their time to

the project. Thirdly, applicants not selected by some, or even any, of the parallel systems would nevertheless have to be employed by the organization. There is no point in following up successful applicants if you do not know how 'unsuccessful' applicants would have fared in the same jobs. Few organizations would be willing to offer contracts of employment to their rejects even for the sake of science!

As the definitive study cannot easily be done it is necessary to put together odd jigsaw pieces of research to build up a picture of what is likely to be the best and most practical selection system. The first question must be about the *reliability* of the selection interview. Reliability is the technical word which applies to the interviewer's ability to measure his own performance. *Inter-rater reliability* is a special form of this and refers to the extent to which two or more interviewers will assess an interviewee's performance in the same way. Those who have studied this usually tape-record an interview and play it back to several selectors who rate it independently. Some studies have found very poor inter-rater agreement under these conditions. However, others (Wagner 1949; Bingham, *et al.* 1959) have shown that a structured marking scheme improves the inter-rater reliability quite dramatically. Bass (1951) has shown that the index of reliability under these conditions can be as high as 0.76 (1 = perfect agreement). *Intra-rater reliability* refers to the extent to which an interviewer will award the same marks to an interview performance he has already marked some time before. In some studies interviewers are asked to re-interview candidates whom they have already interviewed. In others, they listen to tapes of interviews they have conducted in the past. Again, difference is found between marking schemes which are unstructured and those which are structured. (Shaw 1952, Pashalian and Crissy 1953, Anderson 1954). Structured marking schemes produce perfectly acceptable indexes of intra-rater reliability (although the interviewers capacity to remember this previous mark is hard to estimate here).

The demonstration of such high-reliability figures is important, because if an interview does not reliably measure itself it can scarcely be *valid*. *Validity* refers to the extent to which an interview measures whatever it is supposed to measure. One way of assessing validity is to compare the interviewer's measurement with some independent external measure of the same characteristic. Supposing an interviewer were expected to measure a candidate's 'intelligence'. The interviewer's rating could be compared with the candidate's performance on a standard intelligence test. This kind of measure is called an index of *concurrent validity*. The main problem with making such a measurement is that it always produces an ambiguous result. The index of validity may not be high, either because the interview is a poor means of measurement or because the criterion against which it is being measured (e.g. the intelligence test) is not very good. This is known as the *paradox of validity* (Heim 1970). For whatever reason, the concurrent validity of selection interviews is not usually very good (Ulrich and Trumbo 1965).

Another, and stronger, form of validity has also been measured. This is the index of the interviewer's capacity to predict a candidate's performance some months or even years hence and is known as *predictive validity*. Again, the external criteria against which the interview is measured may be questioned. These criteria may be managers' ratings of an individual after he or she has been in a job for some time. However, managers may take a narrow view of job success. In studies made of University or College entrance interviews, the criterion may be success in Final examinations. However, there is more to a successful University career than just doing well in examinations. Despite these problems attempts have been made to measure the predictive validity of interviews. Results suggest that judgements from a single interview are not as good at predicting job or examination success as are most standard tests of mental abilities (i.e. 'intelligence' tests). However, in cases where a team of interviewers were working independently and then came together to make a final group decision the combined assessment was as good as or better than the result obtained from objective paper-and-pencil tests (Rundquist 1947; Trankell 1959).

Another way of looking at interview validity is to ask how much more accurate does the interview make selection compared with selection from past records, tests and other materials in the dossier — to which interviewers would normally have access. In fact the interview does add substantially to follow-up correlations. The interview can complement other data, such as IQ scores, by obtaining information not easily gathered in other ways, such as style of interpersonal behaviour and motivation (Ulrich and Trumbo 1965).

There is also some positive evidence favouring the structured interview. Anderson (1954) reports a study in which its validity was +0.51. This was 0.27–0.37 higher than the predictive validity of non-interview data including school test grades, short-answer tests and essay tests. Thus a structured interview can give a better prediction than any other data commonly available to the selector. Some interviewers make consistently better predictions than others. Vernon and Parry (1949) report that one woman Naval Officer on an interviewing board made assessments which proved significantly better than the predictions from the scores gained by candidates on a battery of objective tests.

Taken together, the research on the reliability and validity of selection interview methods suggests that the best system will incorporate the following features:

(1) A structured system of judgement or rating.
(2) A team of interviewers working independently within the same guidelines and combining information after all interviews have been completed.

Another main feature should be added to these. The best predictor of individual behaviour appears to be behaviour in a similar situation in the past (Mischel 1968). The validity of that idea was discussed at the beginning of the present chapter. Thus the third important feature to be incorporated in a selection interviewing system must be:

(3) A concentration by the interviewer on the interviewee's past behaviour in situations similar to those of the prospective job.

If selection according to this method seems time consuming and difficult, it is as well to remember that the decision about whether or not to hire someone is one of the most important 'purchasing' decisions many managers and administrators have to make. The cost of employing the wrong person is probably at least two years of that person's salary plus any expense caused by his or her inefficiency at the job. In most organizations equipment or materials to that value is bought only after careful research and long hours spent using hard evidence to convince others of the potential value of such a purchase. Yet the idea of lavishing the same amount of time on an efficient system for selection of manpower is often thought to be unnecessary.

Acknowledgement

The help given by John Newton, Training Manager, Alcan Metal Centres Ltd, during the preparation of this chapter is much appreciated. The specimen selection forms are based on some currently used by that company.

References

Anderson, R. C. (1954). The guided interview as an evaluative instrument. *J. Educ. Res.* 48, 203–9.

Argyle, M. and Kendon, A. (1967). The experimental analysis of social performance. *In* Berkowitz, L. (ed.). *Advances in Experimental Social Psychology*. London: Academic Press, vol. 3, pp. 55–98.

Bass, B. M. (1951). Situational tests: I. Individual interviews compared with leaderless group discussions. *Educ. Psychol. Measurement* 11, 67–75.

Beattie, G. W. (1978). Floor apportionment and gaze in conversational dyads. *Br. J. Soc. Clin. Psychol.* 17, 7–16.

Bingham, W. V. D., Moore, B. V. and Gustad, J. W. (1959). *How to interview*. New York: Harper and Sons, 4th edn.

Cook, M. (1979). *Perceiving Others*. London and New York: Methuen.

Cronbach, L. J. (1955). Processes affecting scores on 'understanding of others' and 'assumed similarity'. *Psychol. Bull.* 52, 177–93.

Darwin, C. (1872). *The Expression of the Emotions in Man and Animals*. London: Murray.

Duncan, S. R., Jr. Strategy signals in face-to-face interaction. *J. Pers. Soc. Psychol.* 37, 301–3.

Ekman, P. and Friesen, W. V. (1971). Constants across cultures in the face and emotion. *J. Pers. Social Psychol.* 17, 124–9.

Forbes, R. J. and Jackson, P. R. (1980). Non-verbal behaviour and the outcome of selection interviews. *J. Occupl. Psychol.* 53, 65–72.

Gorman, C., Clover, W. H. and Doherty, M. E. (1978). Can we learn anything about interviewing real people from 'interviews' of paper people? Two studies of the external validity of a paradigm. *Organ. Behav. Hum. Perf.* 22, 165–92.

Hall, E. T. (1966). *The Hidden Dimension*. New York: Doubleday.

Heim, A. (1970). *Intelligence and Personality: their Assessment and Relationship*. Harmondsworth: Penguin Books.

Hollandsworth, J. G. *et al.* (1979). Relative contributions of verbal, articulative, and nonverbal communication to employment decisions in the job interview setting. *Personnel Psychol.* 32, 359–67.

Jackson, D. N., Peacock, A. C. and Smith, J. P. (1980). Impressions of personality in the employment interview. *J. Pers. Soc. Psychol.* 39, 294–307.

Joiner, D. (1976). Social ritual and architectural space. *In* Proshansky, H. M., Ittelson, W. H. and Rivlin, L. G. (eds). *Environmental Psychology*. New York: Holt, Rinehart and Winston, 2nd edn.

Jones, E. E. and Nisbett, R. E. (1971). The actor and the observer: divergent perceptions of the causes of behavior. *In* Jones, E. E. *et al.* (eds). *Attribution: Perceiving the Causes of Behavior*. Morristown, N.J.: General Learning Press.

Keenan, A. (1976). Effects of the non-verbal behaviour of interviewer on candidates' performance. *J. Occupl. Psychol.* 49, 171–6.

—————— (1977). Some relationships between interviewers' personal feelings about candidates and their general evaluation of them. *J. Occupl. Psychol.* 50, 275–83.

_____ (1978). Selection interview outcomes in relation to interviewer training and experience. *J. Soc. Psychol.* 106, 249–60.

Kendon, A. (1967). Some functions of gaze-direction in social interaction. *Acta Psychol.* 26, 1–47.

Korda, M. (1976). *Power in the Office.* London: Weidenfeld and Nicolson.

McHenry, R. (1972). *An Analysis of the Ability to Form Impressions of Other Persons.* D.Phil. thesis, University of Oxford.

Mehrabian, A. (1968). Relationship of attitude to seated posture orientation and distance. *J. Pers. Soc. Psychol.* 10, 26–30.

Mischel, W. (1968). *Personality and Assessment.* New York: Wiley.

Pashalian, S. and Crissy, W. J. E. (1953). The interview: IV. The reliability and validity of the assessment interview as a screening and selection technique in the submarine service. *MLR Rep.*, No. 216, XII (1), January.

Rundqvist, E. A. (1947). Development of an interview for selection purposes. *In* Kelly, G. A. (ed.). *New Methods in Applied Psychology.* College Park, Md.: University of Maryland Press, pp. 85–95.

Rutter, D. R., Stephenson, G. M., Ayling, K. and White, P. A. (1978). The tuning of looks in dyadic conversation. *Br. J. Soc. Clin. Psychol.* 17, 17–22.

Schmitt, N. (1976). Social and situational determinants of interview decision: implications for the employment interview. *Personnel Psychol.* 29, 79–101.

Schneider, D. J., Hastorf, A. H. and Ellsworth, P. C. (1979). *Person Perception.* Reading, Mass.: Addison-Wesley.

Shaw, J. (1952). The function of the interview in determining fitness for teacher training. *J. Educ. Res.* 45, 667–81.

Siegman, A. F. (1976). Do noncontingent interviewer mm-hmms facilitate interviewee productivity? *J. Consult. Clin. Psychol.* 44, 171–82.

Sommer, R. (1959). Studies in personal space. *Sociometry* 22, 247–60.

Trankell, A. (1959). The psychologist as an instrument of prediction. *J. Appl. Psychol.* 43, 170–5.

Ulrich, L. and Trumbo, D. (1965). The selection interview since 1949. *Psychol. Bull.* 63, 110–16.

Vernon, P. E. and Parry, J. B. (1949). *Personnel Selection in the British Forces.* University of London Press.

Wagner, R. (1949). The employment interview: a critical summary. *Personnel Psychol.* 2, 17–46.

Webster, E. C. (1964). *Decision-Making in the Employment Interview.* Montreal: McGill University Press.

Wiggins, J. S. (1973). *Personality and Prediction.* London: Wiley.

2 Skills in the research interview

MICHAEL BRENNER

The idea of measurement in the research interview

The research interview, which is the major data-collection instrument in survey research, is characterized best by its purpose. Cannell and Kahn (1968, p. 527) have summarized a widely shared view: the research interview 'can be defined as a two-person conversation, initiated by the interviewer for the specific purpose of obtaining research-relevant information and focused by him on content specified by research objectives of systematic description, prediction, or explanation'. In contrast to other uses of the interview, the research interview is thus not used just for information gathering; its use is constrained by a particular form of information gathering: *measurement*. As Cannell and Kahn (ibid., p. 530) have pointed out:

> 'The interview is one part, and a crucial one, in the measurement process as it is conducted in much of social research. As such the use of the interview is subject to the laws of measurement; it can be properly judged by the standards of measurement, and it suffers from the limitations of all measurement processes in degrees peculiar to itself.'

The key concepts in determining adequacy of measurement are, of course, validity and reliability. Validity is commonly defined as the extent to which a measure accurately reflects the phenomenon it purports to measure. A measure is invalid to the extent that it measures something more than, or less than, it purports to measure. Reliability refers to the extent to which a measure is likely to yield a consistent score or result; that is 'independent but comparable measures of the same object (or attitude, or whatever) should give similar results (provided, of course, that there is no reason to believe that the object being measured has in fact changed between the two measurements)'. (Selltitz, *et al.* 1962, p. 148.) Validity and reliability of measurement in the research interview stand or fall with the adequacy or inadequacy of three factors: the questionnaire, the respondent and the interviewer. Research interviewing usually uses a structured questionnaire. This means that all questions and instruments to be used in data collection, such as prompt cards, are designed before interviewing and are grouped together to form a particular sequence. Issues of questionnaire design are beyond the scope of this chapter (see Oppenheim 1968); the major objectives in devising a structured questionnaire are, however, straightforward. All questions must be designed so that they use particular sets of stimulus material which are then used uniformly in all the interviews of a data-collection programme. The questions must be grouped together in only one particular sequence so that the order in which the questions are asked is uniform. Also, the content of the questions must be accessible to the respondents.

While the measures to be used in the interview are all listed in, or associated with, the questionnaire, it is only by means of interviewer—respondent interaction that measurement can take place. The fact that measurement in the research interview relies on the medium of social interaction imposes particular constraints on respondent and interviewer.

The respondent must be competent and willing to act in the interview as desired by the researcher. There are three factors which influence the respondent's adequacy of performance in the interview: 'These are *accessibility* of the required data to the respondent, *cognition* or understanding by the respondent of his role and the informational transaction required of him, and *motivation* of the respondent to take the role and fulfil its requirements.' (Cannell and Kahn 1968, p. 535.) Information sought from the respondent may be inaccessible to him because he cannot recall it or has repressed it because of the social desirability of the question content (see Phillips 1973) or other reasons, or because of problems of linguistic communication — the respondent does not understand the question adequately, has problems in forming an adequate answer, or both. Issues of cognition and motivation refer to the requirement that adequate information reporting in the research interview relies on the respondent's conscious understanding and active sharing of the demands made of him.

The interviewing procedures used must not affect the answering process of the interviewer, other than to facilitate the provision of adequate answers, so that validity of measurement is maximized. To secure reliability of data collection all interviewing procedures must be equivalent, which necessitates the standardization of interviewing technique.

When the questionnaire is well designed and respondents and interviewers co-operate in measurement as desired by the researcher, the measures obtained in a data collection may be regarded as adequate and equivalent, which is — of course — a prerequisite for the statistical manipulations used in data analysis. However, in research interviewing it has often proved difficult to achieve adequate measurement fully because of various forms of bias arising during the interview (see Hyman, *et al.* 1954; Sudman and Bradburn 1974). All sources of bias are invariably related to the questionnaire, the respondent and the interviewer as well as to the interactions between them. I cannot attempt a general review of sources of bias in the research interview here; instead, I shall briefly consider the research on biasing interviewer effects as it allowed the development of adequate interviewing technique.

Studies carried out early in the history of survey research indicated that the interviewer's presence in the interview can produce large biasing effects on answers (see Rice 1929; Blankenship 1940). The first large-scale investigation of undesirable interviewer effects was carried out by Hyman, *et al.* (1954) at the National Opinion Research Center in Chicago. Using a wide range of phenomenological, experimental and other data, Hyman and his colleagues found two sources of bias deriving from the interviewer. First, they elaborated the tacit operation of interviewer expectancies on answers: 'how the interviewer enters the situation with certain attitudes and beliefs, which operate to affect his perception of the respondent, his judgement of the response, and other relevant aspects of this behavior.' (Hyman, *et al.* 1954, p. 138.) Second, they stressed the importance of inadequate interviewer performance as a source of bias in terms of cheating and of asking, probing and recording errors (ibid., pp. 225–74).

The framework for the study of biasing interviewer effects first invented and systematically explored by Hyman, *et al.* (1954) was considerably differentiated by further research. As regards bias arising from expectational processes, Kahn and Cannell (1957), for example, noted three general sources of its origin. The first relates to the *content* of the interview:

> 'If an interviewer of strong pacifist persuasions is assigned the task of conducting an interview on attitudes toward the hydrogen bomb, it is entirely possible that his attitudes toward force as a means of settling international disputes will influence the end product of the interview.'
> (ibid., p. 185.)

The second relates to *social perceptions* operating in the interview. That is, age, sex, race and socioeconomic status of the respondent may relate to attitudes or stereotypes which the interviewer holds. Accordingly, the interviewer may make the respondent realize that he expects him to evaluate certain topics in particular ways. The respondent may then answer in terms of a self-fulfilling prophecy; that is, the responses merely reflect the expectations of 'right' answers conveyed to the respondent. The third relates to the *behaviour* of either participant during the course of the interview. As Kahn and Cannell (1957, p. 186) have put it:

> 'The interviewer or respondent finds, in some remark or expressed attitude of the other person, a stimulus which evokes certain attitudes and expectations on his own part and leads him to "fill in" a more complete picture of the person. If a respondent tells an interviewer that he votes for a certain political party, or that he reads a certain newspaper or belongs to a certain political party, the interviewer will expect to discover — and may even assume he is discovering — a whole set of attitudes that he associates with the reported behavior or attribute. The remainder of the interview may reflect these expectations and attitudes of the interviewer, even if the respondent himself fails to fit the stereotype so neatly.'

There exist many studies, most of which are early and have been summarized by Hyman, *et al.* (1954, pp. 225–74) and Kahn and Cannell (1957, pp. 187–93), which have demonstrated the biasing effects of inadequate interviewer performance. Many of these effects are attributable to failures in the interviewer's elementary skills competence. Evans (1961), for example, reported considerable cheating of interviewers in a particular survey. Other forms of bias often result from the interviewer's failure to administer the questions

as required (which alters the stimulus conditions), to prob
adequately (which means that a definite direction of answering
suggested to the respondent or irrelevant information is obtained
and to record the answers accurately (which results in misrepresen
ation of the respondent's answers), besides other issues. Guest (1947)
for example, in a field experiment with fifteen college students inter
viewing one predetermined respondent about his attitudes toward
psychologists, detected 279 classifiable mistakes due to interviewe
incompetence among which twenty-nine were mistakes in reading th
questions, 136 incorrect recording and ninety-seven incorrect
probing. The distribution of mistakes by interviewers showed con
siderable inter-interviewer variance in the types and frequency of
mistakes made. The pattern of findings reported first by Guest (1947)
was replicated in many other studies (see, for example, Hyman, *et al.*
1954, pp. 236–43, or, most recently, Brenner 1981*b*).

While it has often been shown that answers are biased because of
interviewer incompetence, it has also been demonstrated that a
adequate control over the interviewer's performance results in a hig
level of reliability of the interviewer's actions in the interview
Marquis and Cannell (1969), for example, conducted a study of inter
viewer–respondent interaction based on 181 tape-recorded interview
which were carried out by four particularly well-trained interviewer
on employment matters. The findings revealed a low level of inter
viewer mistakes. Of 49 845 coded interviewer actions 1482 (3.0 pe
cent) referred to incorrect question asking, 1269 (2.5 per cent) t
directive probing, 861 (1.7 per cent) to irrelevant conversation an
88 (0.2 per cent) to suggesting an answer. This result demonstrate
the beneficial effects of careful interviewer training on interviewe
performance.

Several studies indicate that an adequate control over the inter
viewer's performance is not only likely to secure a high level of
reliability but also influences positively the level of valid informatio
reporting. Marquis (1967), for example, conducted a study which
used a reinforcement technique with the aim of inducing greate
respondent effort in recalling health information. One group of
interviewers was required to use reinforcing statements ('I see'
'Thank you', 'That's important') after each reported instance of
morbidity, extra explanatory words in the questioning and facial an
postural cues such as smiles and hand gestures, while a control grou
was asked not to use any reinforcing statements and to administer th

questionnaire without using extra words and non-verbal (NV) cues. The reinforcement technique produced an increase in the responses:

> 'the experimental interview obtained significantly greater frequencies of reporting two or three major classes of health events (health conditions and health symptoms), both known to be under-reported at very high rates in the usual household interview about health.'
> (Marquis 1967, pp. 85–6.)

In a more refined study, Marquis and Cannell (1971) used three inter-viewing techniques to investigate the effects of interviewer behaviour on recalling health information. While interviewers were trained in the reinforcement and control interviewing procedures used by Marquis (1967), a third group of interviewers was required to administer a 'sensitizing process' in the form of reading a symptoms list at the beginning of the interview in addition to the control inter-viewing procedures. Again, it was found that more symptoms, con-ditions and illnesses were reported when the reinforcement technique was used: 'In this study, the reinforcement technique elicited about 29 per cent more reports of symptoms than did the sensitization tech-nique.' (Marquis and Cannell 1971, p. 19.) Reading out the symptoms list at the beginning of the interview, however, had only a minimal positive effect on the reporting of health data in comparison with the control interviewing.

While the research on biasing interviewer effects has allowed a high degree of adequate control over the interviewer's performance, this does not mean that other sources of bias, such as problems of language and communication or the lack of willingness on the part of the respondent to report adequately, are equally well controlled. Thus adequate interviewing technique is only one part, if a crucial one, of the conditions which must be met to secure adequate measurement in the research interview. This implies, of course, that even under conditions of excellent interviewer performance the validity and reliability of measurement may be low.

Control over the interviewer's skills competence comprises three areas of activity. First, the skills to be used by the interviewer must be *developed*. Considerable research effort has gone into the develop-ment aspect of interviewing skills, and it is now quite clear what inter-viewers must do to gather information adequately (see, in particular, Gordon 1975; *Interviewer's Manual* 1976). Second, interviewers must be *trained* in the appropriate use of the skills of research interviewing.

Various suggestions for the general outline of trainings have been made (see, in particular, Moser and Kalton 1975; Hoinville, *et al* 1978), but both the content and the structure of effective interviewer training remain so far quite unspecified. Third, for any particular data-collection programme there must be an *assessment* of the degrees to which the interaction between interviewer and respondent has succeeded or has failed in accomplishing adequate data collection. The aspect of assessment of interviewing performance, although it allows a direct investigation of the various forms of bias arising from interviewer–respondent interaction, has so far largely been ignored with only a few exceptions (see, in particular, Cannell, *et al.* 1975). will now turn to a discussion of these three aspects of control over the interviewer's performance.

Rules in research interviewing

Research into the biasing effects of inadequate interviewer performance, together with the wealth of practical experience in research interviewing accumulated over the last two decades, has led to the development of particular 'Do's and Don'ts of interviewing' (Smith 1972, pp. 59–61; Atkinson 1971; *Interviewer's Manual* 1976), or rules of interviewing, which when followed enable the interviewer to gather information, usually with maximum reliability, accuracy and validity. I shall consider here only the rules to be followed by the interviewer while administering the questionnaire to the respondent (for further discussion see Denzin 1970, pp. 132–9; Brenner 1978). These rules are listed in Table 2.1, together with the reasons for their existence.

Table 2.1   Rules in research interviewing (I = interviewer; R = respondent)

Rule	Reason
(a) *Rules for asking questions*	
I must read the questions as they are worded in the questionnaire.	To avoid alteration of question content and/or question form.
I must read slowly.	To provide full opportunity for *R* to understand the question.
I must use the correct intonation and stress patterns.	To facilitate *R*'s adequate understanding of question content.
I must ask the questions in the order in which they are presented in the questionnaire.	To avoid alteration of the intended sequence of questions.

Table 2.1 — *cont.*

Rule	Reason
I must ask every question that applies to *R*.	To avoid missing data.
I must use prompt cards and other instruments where required.	To obtain the kinds of answers desired by the researcher.

(b) *Rules for acting in answering*

I must record the answer verbatim.	To avoid misrepresentation of *R*'s answer.
I must not indicate verbally or non-verbally approval, disapproval, shock, surprise or any other emotion at any answer given by *R*; that is, *I* must display a non-judgemental attitude.	To avoid effect of *I*'s affective evaluation of answers on *R*.
I must not give directive information about question content.	To avoid effect of *I*'s evaluation of matters on *R*.
I must not answer for *R*.	To avoid bias arising from *I*'s judgement of the adequate answer.
I must express an interest in the answers provided by *R*.	To keep *R* motivated to perform his role adequately.
I must only probe non-directively.	To avoid implying or suggesting a particular answer or range of answers.
I must make sure that he has correctly understood an answer and that it is adequate.	To avoid misrepresentation arising from selective comprehension of *R*'s answer and the acceptance of an inadequate answer.
I must not seek or give unrelated information.	To avoid distracting *R* from answering the questions.
I must express appreciation of *R*'s adequate performance, at least occasionally.	To reinforce *R* in his role performance.

(c) *Rules for dealing with respondent problems*

When *R* requires a particular action of *I* (a question, a probe, an instruction, a clarification) to be repeated *I* must repeat the action.	As *R* indicates comprehension difficulties *I* must repeat.

Table 2.1—*cont.*

Rule	Reason
When R asks for clarification I must clarify non-directively and in accordance with the question objectives.	As R indicates problems of understanding what is meant by a question or probe or instruction I must clarify; this must be done non-directively and in accordance with the question objectives to avoid alteration of question content and/or form.
When R gives an inadequate answer I must act non-directively (probe, repeat question, repeat instruction, clarify) towards obtaining an adequate answer.	To avoid inadequate answers being accepted.
When R definitely refuses to answer a question (in the form 'I'm not going to answer that question') I must accept the refusal; otherwise, I must act non-directively (repeat question, repeat instruction, clarify) towards obtaining an adequate answer.	R has the right to refuse to answer questions; I may try to overcome a 'half-hearted' refusal if R finds the question or range of answers inapplicable to his situation but is otherwise co-operative.

Inspection of these rules shows that the interviewer's performance uses not only verbal but also non-verbal skills in six specific areas of interviewer—respondent interaction: questioning, answering, repetitions, clarifications, inadequate answering and refusals. Before outlining specific action repertoires for any of these areas I shall first consider some general social psychological aspects of research interviewing.

General social psychological aspects of research interviewing

Research interviewing uses actions on two levels, that of interaction between interviewer and respondent; and embedded in this, communication by *only* the respondent about issues. In the (rough) transcript of a question—answer sequence in Table 2.2 I have marked these action levels. Action descriptions are also provided, from which it is clear that the respondent's communication about issues is part of a particularly rule-constrained interaction structure between interviewer and respondent where finally, after some considerable problem solving, the desired answer appears.

Table 2.2 Question–answer sequence (I = interviewer; R = respondent).

Action	Action level	Action description
I: And how satisfied are you with the physical surroundings of the Upper Afan? (pause). It's that card here (pause). Satisfied? (pause). Very satisfied? (pause). Can you take these answers because, otherwise, you know.	Interaction	I reads question as required, uses card as required and clarifies.
R: Yes (pause). Say that question again.	Interaction	R gives feedback and requests repetition of question.
I: Yes (pause). How satisfied are you with the physical surroundings of the Upper Afan? (pause).	Interaction	I gives feedback and repeats question.
R: What do you mean?	Interaction	R requests clarification.
I: Well, um (pause), like the scenery, landscape and so on.	Interaction	I clarifies.
R: Good scenery isn't it now.	Communication	R answers inadequately.
I: (inaudible).	Interaction	
R: Fine by the sea isn't it?	Communication	R answers inadequately.
I: No; we want the physical surroundings.	Interaction	I clarifies.
R: Yes, that's not too difficult. You couldn't find it better anywhere else.	Interaction Communication	R gives feedback and answers inadequately.
I: So (pause) what would your answer be?	Interaction	I probes.
R: Satisfied.	Communication	R answers adequately.
I: Satisfied.	Interaction	I repeats the answer.

The verbal action dimension, as transcribed above, is in itself embedded in an NV action context. In all, besides the verbal component of action, five NV components must be considered: the environmental organization of the interview, its chronemic (or time) organization, and paralinguistic, spatial and kinesic behaviour, of which not all are equally relevant to interviewing skills.

Issues related to the environmental organization of the interview setting are often unimportant, because interviewer and respondent usually organize their relative positions in the setting so that each feels comfortable. However, it may be necessary, as Wilson, *et al.* (1978) found when conducting research interviews with members of local authority services departments, to make particular arrangements for a desirable interviewing environment. In their case, these included moving the respondent from his busy office into a quiet room, diverting telephone calls and preventing other contacts for the interview period, among other things.

Similarly, the spatial aspect of action can often be ignored, because seating is usually arranged so that distance and orientation between interviewer and respondent are convenient and culturally acceptable. Spatial aspects of action will need consideration, however, when interviewer and respondent will probably apply different organizing principles to their physical conduct.

While the environmental and spatial aspects of action usually need little consideration, this is not so with the other NV components of action. There is an explicit chronemic rule; namely, that questions must be asked slowly. As a general requirement, the interviewer should also conduct the interview quite slowly, as a deliberate pace of interviewing provides time not only for the respondent to answer or to raise problems but also for the interviewer to write down an answer accurately or to prepare himself well for the next question. 'High speed' interviewing is stressful for interviewer and respondent in that both feel continuously as if they are 'running out of time'; also, recording errors and other mistakes will probably occur. Contrastingly, if the interviewer uses too much time this may lead to irritation or embarrassment on the part of the respondent, as the interviewer may be experienced as incompetent or as 'wasting his time'. Besides the issue of the adequate pacing of actions, silence in itself may be used as action. Probing may be done just silently; so may waiting for an answer to appear (see Gordon 1975, pp. 376–8).

The kinetic component of action refers to bodily action channels other than speech, such as facial expression, gaze, hand and foot movements and other body movements. Research interviewing requires a considerable kinesic component, such as the need for the interviewer to use prompt cards, to display a non-judgemental attitude, express interest in the respondent's answers and appreciate his performance.

While the use of cards consists of a sequence of simple hand movements, the other rules require more kinesic effort. A non-judgemental attitude may be expressed by carefully concealing the expression of undesirable affect in the face. As masking emotions in the face is not ordinarily done by people, at least not all the time, the interviewer must learn to monitor and, if necessary, adjust her facial expression. Interest in the respondent may be shown kinesically with head-nods and gaze directed towards the respondent in the form of an 'expectant gaze'. Appreciation may be given by a combination of smiling, nodding and looking towards the respondent and a linguistic action such as 'Fine' or 'Mmhmm'. Finally, adequate writing skills may develop only with considerable practice as verbatim recording is required. As kinesic actions are particularly useful for punctuating, emphasizing and synchronizing linguistic actions, providing feedback and signalling attention, the kinesic action layer must receive particular consideration in interviewer training.

There are two relevant dimensions to the paralinguistic component (see Argyle 1975, p. 345). The first relates to voice quality, in particular tone and type of voice, the former communicating emotion and the latter information about social and personality characteristics of the speaker. The interviewer must control, as with kinesic actions, the degrees to which his linguistic performance transmits emotions which are likely to affect his non-judgemental position. Little can be done, however, about the type of voice of the interviewer, once selected. As certain types of voice, such as social and regional accents, carry specific social meanings within particular groups it may be necessary to consider whether these meanings can have undesirable effects on an interview programme.

The second paralinguistic dimension relates to the vocal features of speech, in particular pitch and stress. The interviewer must ask the questions using the appropriate stress and intonation patterns so that their intended meaning is most likely to be communicated. In interviewer training it may be useful to role play question asking; for example, by using questions represented in the form of tone-units (see Brazil 1975) within which the words to be emphasized are marked. Instead of providing a question in the form 'How many relatives do you have living in this area whom you see at least once a fortnight?' it may be better to write the question in tone-unit form, such as '//How many *relatives*//do *you* have//*living* in this area//whom you *see*//at least *once* a fortnight?//'. This provides an explicit format for the paralinguistic communication of meaning in question asking.

An action repertoire for the interviewer

For the interviewer to follow the rules of interviewing effectively he
must possess a repertoire of actions which enables him to perform
adequately. In designing such a repertoire several issues need to be
considered.

First, such a repertoire must include not only all the action
necessary and sufficient for adequate interviewing but also sequence
of action suitable for solving the kinds of problems that can arise
in interviewer—respondent interaction and for adequate synchron
ization of the three main areas of skill for the interviewer: the
administration of the questionnaire, effective interaction with the
respondent during answering and in problem solving, and recording
the answers.

Second, as questions vary in terms of the form of answers desired
the action sequences must be designed to take this into account
There are two main types of question, those inviting 'closed' answers
that is, answers which are predefined in some sense and from among
which the respondent has to make a choice; and those inviting 'open
answers, that is, answers which are undetermined in structure and
content.

Third, as indicated already, question—answer sequences vary in
terms of problems created by respondents. There are two main type
of question—answer sequences; those which use an unproblemati
elementary structure where the question is followed by a set of action
by respondent and interviewer required in the communication and
recording of an adequate answer, and those which entail one or more
problems where question—answer sequences are complicated by
requests for repetition of the interviewer's actions, clarifications
inadequate answers and refusals.

Fourth, the action sequences must be designed so that three
requirements can be fulfilled: the respondent must have sufficient
time to comprehend the question and the task required of him in
answering, the interviewer must be able to fully assess the adequacy of
an answer and he must be able to record the answer fully and to
establish whether the information recorded is correct.

Fifth, the design of action sequences must be based on a consider
ation of the specific social interactional characteristics of interview
ing. The verbal and non-verbal components of action must be
integrated and the particular sequences of action designed must allow

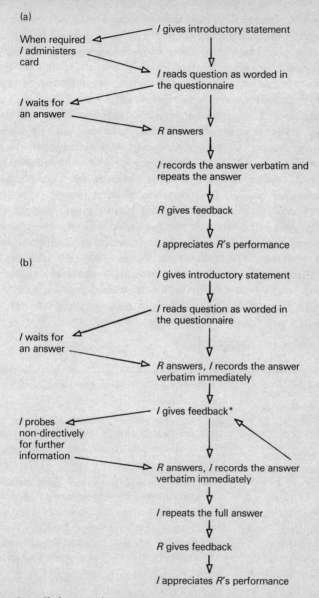

(a)

I gives introductory statement

When required
I administers
card

I reads question as worded in
the questionnaire

I waits for
an answer

R answers

I records the answer verbatim and
repeats the answer

R gives feedback

I appreciates R's performance

(b)

I gives introductory statement

I reads question as worded in
the questionnaire

I waits for
an answer

R answers, I records the answer
verbatim immediately

I gives feedback*

I probes
non-directively
for further
information

R answers, I records the answer
verbatim immediately

I repeats the full answer

R gives feedback

I appreciates R's performance

Figure 2.1 Skeleton action structure used (a) 'closed' questions and (b) in 'open' questions. *I* = interview; *R* = respondent. At this point in the sequence the interviewer may engage in a cycle of feedback and probing until the question objectives have been met.

easy monitoring for the interviewer so that he is able to control and manage the situation throughout. The action sequences must also be natural in that the organization and appearance of actions is experienced by the respondent as 'ordinary'.

In most instances, questions are simply followed by a set of actions which is required in the communication and the recording of an adequate answer. The action sequences proposed below offer just a broad solution to the design of such elementary question–answer sequences. For any particular interview programme, these action sequences may need refinement and alteration; for example, when instruments such as batteries of attitude scales are used or when complex 'open' answers (accounts) are invited. Only the most prominent action characteristics are mentioned here; the more fine-grained kinesic and other NV features are not considered. My discussion will, however, include those aspects of action which will need particular attention in interviewer training. The elementary action structures used in the administration of 'closed' and 'open' questions are represented in Figure 2.1 in their basic skeleton forms. A more detailed view of the action structure associated with 'closed' questions, may take the form shown in Table 2.3.

Table 2.3 Action structures associated with 'closed questions'
(I = interviewer; R = respondent).

Action	Explanation	Example
I gives introductory statement.	I opens the question–answer sequence looking at R.	Now we have a question on employment.
When required, I administers card.	I hands card over to R and explains the procedure so that R can used card as required.	Can you look at this card and give me an answer from the card please?
I reads question as worded in the questionnaire, slowly and meaningfully.	I looks down on the questionnaire and by reading slowly and using the correct intonation and stress patterns provides full opportunity for R to understand the question.	//How *satisfactory*// do *you* think//*job* opportunities//are *generally*// in *this* area?//

Table 2.3—*cont.*

Action	Explanation	Example
I waits for an answer.	*I* looks towards *R* in an expectant manner, thus indicating that it is now *R*'s turn to provide an answer.	
R answers.	*R* provides an adequate answer.	Very unsatisfactory I should think.
I records answer verbatim and repeats answer.	*I* looks down on the questionnaire and records the answer verbatim, then looks up towards *R* and repeats the answer so that *R* can assess whether *I* has understood the answer correctly; this also provides an opportunity for *R* to correct the answer if necessary.	Very unsatisfactory.
R gives feedback.	*R* indicates whether *I* has correctly received the answer or whether changes should be made.	Mmhm.
I appreciates *R*'s performance.	*I* looks at *R*, smiles lightly and indicates that the sequence has been successfully completed, then looks down on the questionnaire to prepare the next question—answer sequence.	Right.

The elementary action structure used in 'open' questions may take the form shown in Table 2.4.

Table 2.4 Action structures, associated with 'open' questions (I = interviewer; R = respondent).

Action	Explanation	Example
I gives introductory statement.	*I* opens the question–answer sequence looking at *R*.	Now.
I reads question as worded in the questionnaire, slowly and meaningfully.	*I* looks down on the questionnaire and by reading slowly and using the correct intonation and stress patterns provides full opportunity for *R* to understand the question.	//*What* would you describe//as the *advantages*//of *living*//in *this* area?//
I waits for an answer.	*I* looks towards *R* in an expectant manner, thus indicating that it is now *R*'s turn to provide an answer.	
R answers, *I* records the answer verbatim immediately.	*R* provides an adequate answer which *I* records verbatim immediately.	It's a friendly community.
I gives feedback.	*I*, looking at *R*, by expressing interest in *R*'s answer encourages *R* to carry on answering.	Yes.
R answers, *I* records the answer verbatim immediately.	*R* carries on answering, and *I* records the information verbatim immediately.	The neighbours often come to see whether they can do anything for me.
I gives feedback.	*I*, looking at *R*, by expressing interest in *R*'s answer encourages *R* to carry on answering.	Mmhm.
I probes non-directively for further information.	When the answer so far provided does not exhaust the question objectives *I* probes non-directively for further information; he looks at *R* while probing.	Are there any other advantages of living in this area?

Table 2.4 — *cont.*

Action	Explanation	Example
R answers, I records the answer verbatim immediately.	R carries on answering, and I records the information verbatim immediately.	Well there is* good entertainment here — the pub, the working men's club — and there is a new sports centre for the younger element in the village.
I repeats the full answer.	I looks first towards R and then down on the questionnaire and repeats the complete answer verbatim so that R can assess whether I has understood the answer correctly; this also provides an opportunity for R to correct the answer if necessary.	I will repeat your answer to make sure that I've understood you correctly; so you said 'It's a friendly community' and 'The neighbours often come to see whether they can do anything for me'; then you said 'There is good entertainment here — the pub, the working men's club — and there is a new sports centre for the younger element in the village'.
R gives feedback.	R indicates whether I has correctly received the answer or whether changes should be made.	Yes.
I appreciates R's performance.	I looks at R, smiles lightly and indicates that the sequence has been successfully completed, then looks down on the questionnaire to prepare the next question—answer sequence.	Fine.

* At this point in the sequence the interviewer may engage in a cycle of feedback and probing until the question objectives have been met.

The action structures often related to 'closed' and 'open' questions will be empirically more complex and varied than outlined above, as the interaction not only relies on the interviewer's plan of action, but also on the respondent's specific decisions for action at any given point during a question—answer sequence. For example, the respondent may often interpunctuate the interviewer's actions by giving feedback in the form of 'Mmhm' or 'Yes' or he may actually correct an answer, which means that the interviewer must now record the corrected answer verbatim and repeat it to the respondent or he may volunteer additional information which the interviewer must ignore. In answering 'open' questions, the respondent may answer too quickly so that the interviewer cannot manage to record the information verbatim. The interviewer may then require the respondent to 'chunk' the answer or to repeat it so that he can write it down. The respondent may change his opinion on matters while answering, which may lead to contradictory information being assembled. If this happens the interviewer must engage in some negotiation of parts of the answer with the respondent (in the form 'You said earlier x and now y; what do you mean?') so that finally a clear picture can be drawn. As it is impossible to outline in advance the detailed patterns of respondent action in elementary question—answer sequences, particular designs of action sequences can only partially be based on the above suggestions and must also consider specific knowledge about the respondent population, which is best gathered through some pre-testing of the interview methods to be used.

The structured action approach to elementary research interviewing, as a general idea, is advantageous in several ways. It provides the interviewer with a clear cognitive 'landscape' of the structure of action used in elementary question—answer sequences; as all moves made by interviewer and respondent have explicit action meanings the interviewer can clearly monitor the particular stage of the sequence with which he is dealing. This means that the likelihood of error occurring in the interviewer's decision making is greatly reduced, which facilitates the reliability of his performance. Also, this approach combines the three main skill areas for the interviewer: the administration of the questionnaire, effective interaction with the respondent during answering and recording the answers, in such a way that these can be realized well within a natural flow of actions. Finally, interviewer training is facilitated, as there are clear requirements for actions to appear in particular forms of sequential organization, which trainees need to learn to meet.

Turning now to problem solving in research interviewing, respondents may decide, for various reasons, to introduce additional action requirements for the interviewer during the course of an otherwise elementry question–answer sequence. First, respondents may ask for an action of the interviewer to be repeated. This is usually done in the form of 'Pardon?' or 'Can you repeat that please?'. Requests for repetition are simply met by the interviewer by restarting the subsequence where the respondent wishes.

Second, the respondent may wish to have issues clarified. Requests for clarification may only be 'rhetorical', as in the following exchange:

Action	*Action description*
I: How much rent do you have to pay for this flat?	*I* asks question as required.
R: How much rent?	*R* requests 'rhetorical' clarification.
I: Yes.	*I* gives feedback.

As indicated, such 'rhetorical' requests are simply met by the interviewer giving feedback. However, the respondent may have serious problems of understanding the question or card or instructions expressed in the form of 'How do you mean?', 'What exactly do you want me to do?', or he may wish to have the meaning of a particular word clarified; for example, 'How do you mean "friends"; we're all friends here; do you mean "personal friends"?' Such serious requests for clarification must be met by the interviewer in a non-directive manner to avoid distortion of the question content, form or both. This is best achieved by means of a repetition of the action which led to the request for clarification. In addition, the interviewer may emphasize the task required by means of further clarification as in the following exchange:

Action	*Action description*
I: What would you describe as the disadvantages of living in this village?	*I* asks question as required.
R: How do you mean 'disadvantages'?	*R* requests clarification of a word meaning.

I: Well it says here: What would you describe as the disadvantages of living in this village? So it's your task to tell me which disadvantages you think there are.	*I* repeats question and gives further clarification.

Third, the respondent may provide an answer which is inadequate in that it does not meet the requirements of the answering task. There are three types of inadequate answers: answers which are semantically unclear or incomprehensible for other reasons, answers which are related to the content of the question but do not meet the form of answering required, and answers which are irrelevant to the task, that is, the respondent gives information independently of the question asked.

The interviewer must attempt to improve semantically unclear or otherwise incomprehensible answers by using non-directive probing, as in the following example:

Action	*Action description*
I: What would be your most important reasons for moving?	*I* asks question as required.
R: Well it's the closure if the closure houses closure isn't it?	*R's* answer is incomprehensible.
I: You said it's the closure; can you tell me about it?	*I* probes non-directively.

Answers which are inadequate because they fail to meet the form of answering required can occur only with 'closed' questions. In such instances the interviewer must repeat the question, the instructions or both and may clarify if necessary, as in the following example:

Action	*Action description*
I: Now can you look at this card; how satisfied are you with the size of your house?	*I* administers the appropriate card and asks question as required.
R: Oh very good.	*R* answers inadequately.

I: Can you look at card 3 please and give me an answer from there; how satisfied are you with the size of your house?	*I* repeats the instruction, clarifies and repeats the question.
R: Oh, very satisfied.	*R* answers adequately.

Respondents may provide information in reply to a question which is irrelevant to the answering task at hand. The interviewer must then repeat the question and/or the instructions; he may also give additional clarification, as in the following example:

Action	*Action description*
I: How often have you discussed moving from this area recently?	*I* asks question as required.
R: Well, as my wife said, if there were a shopping centre near us we would like to stay.	*R* gives irrelevant information.
I: Sorry; this is about the number of your recent discussions about moving from this area; how often have you discussed moving from this area recently?	*I* clarifies and repeats question.
R: Oh, twice recently.	*R* answers adequately.

Finally, and rarely, the respondent may refuse to answer a question. There are two forms of refusal, as previously indicated; the 'definite' refusal where the respondent points out that he will not co-operate at all, which the interviewer must accept, and the 'half-hearted' refusal, where the respondent may raise a number of specific problems which make it impossible or difficult to answer, but is otherwise co-operative. 'Half-hearted' refusals have often to do with the task of answering required with 'closed' questions. For example, respondents may find the predetermined range of answers inapplicable to their situation or the question as a whole may be experienced as unanswerable in the form required. The interviewer must try to over-come a 'half-hearted' refusal by repeating the question and/or the instructions and, if necessary, by giving additional clarification, as in the following exchange:

Action	*Action description*
I: How satisfied are you with the provision of shopping facilities in this area?	*I* asks question as required.
R: Well I can't answer that because there is nothing here is there?	*R* gives 'half-hearted' refusal.
I: Well, we are interested in your satisfaction with the provision of shopping facilities in this area; can you look at card 3, please, and give me an answer from there.	*I* clarifies and repeats the instruction.
R: Very dissatisfied, because there is nothing here.	*R* answers adequately.

The interviewer's repertoire of non-directive problem solving is entirely composed of repetitions, clarifications and probes. Probes may, of course, be used not only in 'repairs' of inadequate answers, but also for the purpose of eliciting further information, as has been pointed out. Table 2.5 summarizes the kinds of adequate probes that are available (see also Atkinson 1971, *Interviewer's Manual* 1976).

Table 2.5 Probes and primary area of application.

Probe	*Primary area of application*
Anything else?	Answering 'open' questions.
Any other reason?	
Can you tell me more about it?	
Why do you feel that way?	
Can you tell me more about your thinking on that?	
Can you explain this a little more?	
In what way?	
Sorry?	'Repairing' inadequate answers
How do you mean?	
What do you mean?	
Can you explain a little more fully what you mean?	
What would your answer be?	
Which answer comes closest to the way that you feel?	

The basic structure of interviewer training

There is general agreement in the literature that interviewers should be trained, but there is no consensus about what methods are most effective and how much training is required; there is considerable diversity of opinion and practice. Cannell and Kahn (1968, pp. 585–6) have observed that the methods used range from a hopeful reliance on written instructions to proposals that, for certain types of anthropological research, interviewers should be psycho-analysed as well as trained. As regards the quantity of training, some organizations give virtually none, 'relying on a quick dismissal of those whose performance fails to come up to scratch; others take the utmost trouble in training their new staff.' (Moser and Kalton 1975, p. 287.)

There exist many descriptions of particular training programmes (see Cannell and Kahn 1968; Smith 1972; Moser and Kalton 1975; Hoinville, *et al.* 1978). These, however, are vague, so that it is impossible to derive from them in detail which paths and methods of skill acquisition should be used. In the notable absence of any published detailed interviewer training programme, I shall give a brief description of my own training method.*

The method was developed incorporating Cannell and Kahn's recommendations for the general outline of interviewer training (1968, p. 586). It incorporates the following objectives:

(1) To provide the trainee with an adequate cognitive representation of the general idea of measurement in the research interview and of the means of social interaction required in adequate interview-ing performance.
(2) To teach the techniques of interviewing effectively.
(3) To provide extensive opportunity for role-playing practice and evaluative feedback with the aim of developing an adequate level of the perceptual and motor skills required in interviewing as well as the ability to critically self-monitor the adequacy of interview-ing performance.
(4) To offer careful evaluation of the adequacy of the training programme in production interviewing.

To be able to provide a training at all, the researcher must first think out clearly the social interactional means needed for adequate data

---

* A videofilm illustrating the method in detail may be obtained from Dr T. D. Wilson, Postgraduate School of Librarianship and Information Science, University of Sheffield.

collection. This means that the design of the questionnaire must be accompanied by devising the appropriate action repertoire to be used in its administration. Pre-testing thus includes establishing not only an instrument which will 'work' satisfactorily with respondents, but also whether the designed action repertoire will secure adequate interviewing performance. After some trial and error a particular action repertoire will emerge which then provides the content for interviewer training. As the action repertoire is organized into action sequences it is not only clear *what* must be done in interviewing, but also *how* the interviewer must organize his conduct in the various question—answer sequences to achieve an adequate level of performance.

As this training method relies on intensive role-playing practice and evaluative feedback as the main means of skills acquisition, the group of trainees must be small. Also, the time required may be quite considerable in comparison with the time usually spent on role playing in initial interviewer training. That is, days of practice may be needed rather than a few hours, if any time is budgeted for role playing at all (see Moser and Kalton 1975, p. 288). To facilitate realism in the role playing the researcher must prepare an interview environment which is likely to reproduce well the interview settings to be encountered in the field. It is also necessary to provide video recording and playback facilities, so that part of the role-playing performances may be filmed for further discussion and scrutiny.

To build up an adequate cognitive representation of the general idea of measurement in the research interview and of the action repertoire required in adequate interviewing the training session starts with a lecture introducing the social organization of measurement in the interview. The rules of interviewing are emphasized and the various action formats occurring in the interview are presented. At this stage the trainees are encouraged to raise any doubts or concerns about the aims and methods of interviewing.

To provide an effective teaching framework for the role-playing part of the training a cumulative learning approach is favoured. This means that the trainees are asked to start role playing the simple tasks first to which more complex action requirements are added at later stages in the training until a final practice level is accomplished.

In practice, the trainees start to role play, using perhaps first the researcher as the respondent, subsequences of actions related to the elementary action sequence of 'closed' questions, such as asking the question as required, looking up towards the respondent and waiting for an answer, answering and recording the answer. Other actions, such as giving an introductory statement, repeating the respondent's answer and appreciating his performance, are subsequently added so that the trainees acquire the ability to perform the overall sequence of actions related to the handling of 'closed' questions. The same procedure is used in learning the elementary question—answer sequence of 'open' questions. The practice covers first single questions and then extends to questions in succession. The performances are discussed and corrected on the spot. Some are filmed and played back for more detailed appraisal. The elementary phase of the training is carried out until it becomes apparent that the trainees master the basic interviewing skills.

The additional action requirements are introduced at the next stage, starting from the straightforward structure of repetitions, then dealing with the forms of clarification and finally moving to inadequate answering and refusals. In the practice of dealing with inadequate answers the importance of adequate evaluation of the respondent's information is stressed to develop the particular perceptual skills that are required in dealing with this kind of respondent action. The additional action requirements, once mastered in their basic forms, are then role played in combination within a question—answer sequence and for questions in succession. Again, feedback and correction is provided, by means of either immediate assessment or video recording.

Towards the end of the training the trainees are required to enact large parts of the questionnaire, including all conditions of interviewer—respondent interaction which are expected in a data-collection programme. Their performances are now filmed at some length and discussed with the aim of eliminating more fine-grained verbal and non-verbal problems in their conduct. I have found that, at the end of the session, the trainees have not only acquired an adequate cognitive representation of desirable research interviewing, action by action, but they have also received intensive cumulative and controlled exposure to the repertoire of action necessary to secure adequate data collection across the many repeated applications of the questionnaire.

How effective is the method of interviewer training I have proposed? So far, although the method has been used successfully in a sizeable research project (Wilson *et al.* 1978), only a little serious evaluation research has been completed. In one study (Brenner 1980) I decided to explore the method's effectiveness by simulating the pre-test stage of a survey project. Eight students, inexperienced in interviewing, participated in the study. The training was carried out over a period of 6 weeks as two 2-hour sessions a week. The study questionnaire included a rather wide range of questions on the respondents' personal circumstances, such as their happiness, number of friends, the environment in which they lived, voting issues and demographic information. All question forms were considered. In all, twenty-six taped interviews were obtained (each interviewer conducted two or more taped interviews) which were submitted to detailed action-by-action analysis. This showed a very low level of interviewer mistakes. Of all questions asked, twelve (1.58 per cent) were appreciably altered. Inadequate answers were accepted in thirty-eight instances (5.03 per cent). Directive probing was used in seven question–answer sequences (0.93 per cent), and answering for the respondent occurred in seventeen sequences (2.24 per cent). These results compare favourably with Marquis and Cannell's (1969) findings summarized above (pp. 32–3). Unfortunately, it is impossible to compare the two evaluation studies for training methods, as Marquis and Cannell (1969) omitted a detailed description of their interviewer training programme from their report.

Assessing the adequacy of interviewing performance

Despite the fact that many survey research organizations invest considerable effort in the training of their interviewers, thus far only a little evaluation research on the effectiveness of the various interviewer training methods used has been conducted. To my knowledge, besides my own work (see Brenner 1980, 1981), there are only two published studies (Marquis and Cannell 1969; Morton-Williams 1979) where the performance of interviewers is systematically evaluated. Cannell, *et al.* (1975) have provided a manual for coding and analysing interviewer behaviour from tape-recordings of interviews; but, as far as I am aware, their particular method has not been used in published work outside the method's place of origin, the Institute of Social Research, Ann Arbor, Michigan, except in the work by Morton-Williams (1979). This lack of application is deplorable, as the general approach to the evaluation of interviewing performance originally advanced by Cannell and his associates is useful in at least three ways. First – and this is a requirement of the interviewer training method proposed above — a direct scrutiny of interviewer—respondent interaction provides the interviewers with detailed objective feedback on the adequacy of their performances in the field, thus replacing the need for any subjective form of assessment — which is usually used — and its undesirable side effects:

'The procedure which is usually used, if in fact any system is used for evaluating actual interviewing performance, is some form of field observation in which a supervisor accompanies an interviewer into the household and observes the interview. This procedure is distasteful to the interviewer and disrupting to the interview. Furthermore, since the supervisor must take notes on his observations it is usual to focus on examples of bad performance rather than a more balanced evaluation. The results thus tend to appear quite negative with resulting poor morale.'

(Cannell, *et al.* 1975, p. 6.)

Second, a direct and systematic assessment of interviewer—respondent interaction highlights the sources and kinds of bias arising from the interaction as well as recording errors. Third, it provides the researcher with evidence of any shortcomings of present interviewing technique. Thus it becomes possible to improve the present repertoire of desirable interviewing actions for future applications.

I shall now briefly outline the basic approach to assessing interviewer—respondent interaction. As a first step it is necessary to obtain actual recordings of interviews. As video equipment is at present cumbersome to operate under field conditions, the researcher will probably settle for audio-recordings only. This means, of course, that only the linguistic component of the interaction can be studied. Today's cassette recorders are easy to operate and they are least likely, of all recording devices, to be experienced as obtrusive in the interview setting, in that once the respondent has agreed to the recording the presence of the recorder is usually not further noticed. It is important to pay some attention to the quality of the cassettes, as some makes are likely to break under field conditions. How many of each interviewer's interviews should be recorded? Usually, 8—10 interviews will probably be sufficient to give a representative insight into an interviewer's performance. In my experience little is gained if more tapes are taken into account.

Once the tapes are returned by the interviewers they can be analysed in detail, that is, action by action as performed by

interviewer and respondent in question—answer sequences, or just i
terms of the occurrence of particular relevant interviewing actions
this being the approach suggested by Cannell, *et al.* (1975). Th
detailed, and time-consuming, action-by-action analysis has th
advantage that not only can frequencies of particular actions b
generated but also the dynamic aspects of interviewer—responden
interaction can be studied using various forms of sequence analysis.
If such an assessment of interviewing performance is planned it mus
be costed carefully, as the coding of the tapes, which depends on th
aims of the analysis, may necessitate considerable effort.

Conclusion

I have attempted to provide an introduction to the design, training an
assessment aspects of research interviewing. Thus the wealth o
knowledge that has been accumulated in the general social psycholog
of the research interview (see, in particular, Gordon 1975) was no
dealt with in any detail. Despite the research effort that has been spen
on the research interview in the past decades, it is true that othe
aspects of research interview practice have not yet received seriou
scientific attention. This applies in particular to interviewer trainin
and to the assessment of interviewing performance. Hopefully, th
expertise available in the more general psychology of training an
assessment (see, for example, Welford 1976; Gagné 1977) will b
assimilated into the field of research interviewing. Today, however, i
is probably safe to end with some wise words from Cannell and Kah
(1968, p. 589): 'Any present-day discussion of interviewing for socia
research should be tentative in tone; the field is in flux.'

* A computer-programme package has been developed at Oxford Polytechnic whic
combines various available methods for frequency and sequence analysis; intereste
parties should contact the author.

References

Argyle, M. (1975). *Bodily Communication*. London: Methuen.
Atkinson, J. (1971). *A Handbook for Interviewers*. London: HMSO.
Blankenship, A. B. (1940). The effect of the interviewer upon the response i
    a public opinion poll. *J. Consult. Psychol.* 4, 134—6.
Brazil, D. (1975). *Discourse Intonation. English Language Research*
    Birmingham University. Discourse Analysis Monographs, No. 1.
Brenner, M. (1978). Interviewing: the social phenomenology of a researc
    instrument. *In* Brenner, M. and Marsh, P (eds). *The Social Contexts o*
    Method. London: Croom Helm.

_____ (1980). Assessing social skills in the research interview. *In* Singleton, W. T. (ed.). *Social Skills*. New York: Plenum Press.

_____ (1981). Patterns of social structure in the research interview. *In* Brenner, M. (ed.). *Social Method and Social Life*. London: Academic Press.

Cannell, Ch. F. and Kahn, R. F. (1968). Interviewing. *In* Lindzey, G. and Aronson, E. (eds). *The Handbook of Social Psychology*. Reading, Mass.: Addison-Wesley, vol. 2.

Cannell, Ch. F., Lawson, S. A. and Hausser, D. L. (1975). *A Technique for Evaluating Interviewer Performance*. Ann Arbor, Mich.: Institute for Social Research.

Denzin, N. K. (1970). *The Research Act*. Chicago: Aldine.

Evans, F. B. (1961). On interviewer cheating. *Pub. Opinion Q.* 25, 126–7.

Gagné, R. M. (1977). *The Conditions of Learning*, New York: Holt, Rinehart and Winston.

Gorden, R. L. (1975). *Interviewing, Strategies, Techniques, and Tactics*, Homewood, Ill.: Dorsey.

Guest, L. (1947). A study of interviewer competence. *Int. J. Opinion Attitude Res.* 1, 17–30.

Hoinville, G., *et al.* (1978). *Survey Research Practice*. London: Heinemann.

Hyman, H. H., *et al.* (1954). *Interviewing in Social Research*. Chicago: University of Chicago Press.

Interviewer's Manual (1976). Ann Arbor, Mich.: Institute for Social Research.

Kahn, R. L. and Cannell, Ch. F. (1957). *The Dynamics of Interviewing*. New York: Wiley.

Marquis, K. H. (1967). *Effects of a Household Interview Technique Based on Social Reinforcement*. Ph.D. thesis, Institute for Social Research, University of Michigan.

Marquis, K. H. and Cannell, Ch. F. (1969). *A Study of Interviewer–Respondent Interaction in the Urban Employment Survey*. Ann Arbor, Mich.: Institute for Social Research.

Marquis, K. H. and Cannell, Ch. F. (1971). *Effect of Some Experimental Interviewing Techniques on Reporting in the Health Interview Survey*. Washington, DC: US Departments of Health, Education and Welfare.

Morton-Williams, J. (1979). The use of 'verbal interaction coding' for evaluating a questionnaire. *Qual. Quant.* 13, 59–75.

Moser, C. A. and Kalton, G. (1975). *Survey Methods in Social Investigation*. London: Heinemann.

Oppenheim, A. N. (1968). *Questionnaire Design and Attitude Measurement*. London: Heinemann.

Phillips, D. L. (1973). *Abandoning Method*. London: Jossey-Bass.

Rice, St. A. (1929). Contagious Bias in the Interview: A Methodological Note. *Am. J. Sociol.* 35, 420–3.

Selltitz, C., *et al.* (1962). *Research Methods in Social Relations*. London: Methuen.

Smith, J. M. (1972). *Interviewing in Social and Market Research*. London: Routledge and Kegan Paul.

Sudman, S. and Bradburn, N. M. (1974). *Response Effects in Surveys.* Chicago: Aldine.

Welford, A. T. (1976). *Skilled Performance: Perceptual and Motor Skills.* Glenview, Ill.: Scott, Foresman and Co.

Wilson, T. D., *et al.* (1978). *Information Needs and Information Services in Local Authority Social Services Departments. Final report to British Library.* University of Sheffield.

3 The social skills of selling

S. E. POPPLETON

Models of sales performance

Each of the three models to be discussed provides a useful framework for examining the social skills of selling. Taken together, they allow a comprehensive and detailed analysis of these social skills.

Argyle and Kendon's social skills model

In essence, this model views social skills as a special class of motor skills (see Fig. 3.1). My concern is with the application of the social skills model to selling.

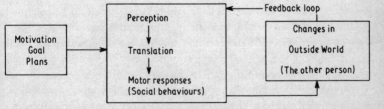

Figure 3.1 Motor skill model (adapted from Argyle and Kendon 1967).

The model in Figure 3.1 refers to only one individual, and it therefore needs to be applied separately both the the salesman* and the customer. The starting point in the model is an individual's motivation and goals, and there is also clear evidence that the presence and strength of certain goals in salesmen are related to their

* The term 'salesman' will be used throughout the text, although the salesman may be a woman.

effectiveness, presumably because of greater motivation to put available social skills into practice.

Plans are concerned with the organization of behaviour so as to achieve the goals which are set. These planning and organizing activities are also related to effectiveness, at least in life-assurance salesmen (Poppleton and Allen in press). Specifically, successful salesmen are more likely than unsuccessful salesman to spend time in planning and organizing their work programmes.

There is also evidence that successful salesmen are more accurate in their perceptions of other people — most notably of potential customers (e.g. Greenberg and Meyer 1964). Specifically, they cite evidence that empathy ('the ability to feel as the other fellow does') is positively related to effectiveness in life-assurance salesmen, a finding which is supported by our work (Poppleton, *et al.* 1977; 1978; Poppleton and Allen 1978). A particularly interesting difference in perception occurred in the attribution of causes of 'rejection' in life-assurance salesmen (Poppleton and Riley, unpublished observations). Less successful salesmen were more likely to attribute being rejected to their own inadequacy as salesmen rather than to the situation. Blaming themselves for failure over a period of time often seemed to bring about a fall in motivation to contact prospective customers. Similarly, Lombard (1955) found that retail salesmen preserve their image of competence in the face of 'rejection' by categorizing customers as 'nasty' or as unpleasant in some other way.

What generally seems to happen in the perception of customers is that the salesman categorizes the customer. For example, Woodward (1960) found that retail saleswomen often used social class as a basis for categorization. Saleswomen tended to feel nervous and apprehensive because of the 'haughty manner' which they perceived as typifying upper middle-class customers. Often, customers were perceived as belonging to more specific categories such as 'peppery colonels' and 'elderly frustrated females'. We also found some evidence of categorization in life-assurance salesmen, and our impression was that successful salesmen tended to arrive more quickly at categorizations which had clear implications for their sales behaviour.

At the level of motor responses, our work has been carried out

* Much of the research data presented here has not been published. Anyone wishing to know about this work in greater detail should write to the author: Department of Economics and Social Studies, The Polytechnic, Wulfruna Street, Wolverhampton.

predominantly at what Trower, *et al.* (1978) term 'higher-order levels' of social skill such as warmth and assertiveness. There is little published evidence on the social skills of salesmen at the level of discrete elements such as gaze, facial expression, voice and verbal influences, hence discussion at this level must remain speculative (e.g. Bonoma and Felder 1977). Those responses which discriminate between successful and unsuccessful salesmen will be described in detail later.

The SPIN* model

Traditional approaches to selling typically advocate that the salesman should focus on customer needs, customer benefits and product features in the sales interview, and usually in that order. For example, Lidstone (1974) states that 'No exercise repeated at regular intervals with each salesman will yield the field sales manager more sales than analysing products in terms of: Customer's needs → Customer benefits → Product features'.

Central to this view is the idea that the sales interaction should be structured such that an analysis of customer's needs should precede that of customer benefits which should in turn precede analysis and discussion of product features.

The SPIN model of Rackham, *et al.* (unpublished observations) can be seen as a development of this traditional approach. SPIN is an abbreviation standing for Situation, Problem, Implication and Need-payoff. Essentially, they argue that the salesman should initially focus on the sales situation and then on the nature of the customer's problem to arrive at the latter's implied need, which the salesman should then help to make explicit.

Their model is based on about a dozen studies which have spanned the past 8 years and they have collected data from over 6500 sales calls by observers trained in a technique of behaviour analysis. This work has its roots in earlier studies conducted by Rackham and others on behaviours in various interactions at work such as appraisal inter-viewing and Union—Management negotiations (see Rackham and Morgan 1977). This work used the analysis of interactions in terms of

* Information on this model and associated research work was supplied by Simon Bailey of the Huthwaite Research Group. Anyone requiring more information on this work should write to the Huthwaite Research Group, Huthwaite Hall, Thurgoland, Sheffield. SPIN and associated materials are copyright and the trademark has been applied for.

eleven general categories of behaviour along with a number of sub-categories for specific types of interaction (e.g. appraisal).

Their category system for the analysis of sales interactions is, however, totally separate. Early research resulted in a thirty-two category system. This has now been reduced to nine categories for practical purposes. Seven of these are concerned with salesman behaviours and two with customer behaviours. The seven salesman behaviours include four which refer to types of questioning corresponding to the SPIN model. Hence, one category refers to 'situation questions', which are essentially background questions. They typically refer to the customer's company, its size, market and various other features. A second category refers to 'problem questions', which probe for the customer's feelings of dissatisfaction with his present situation. The customer behaviours included in the category system are statements of dissatisfaction or of the nature of the problem.

The third category comprises implication questions and seems to be a particularly important area. Questions of this kind are intended to have the effect of highlighting the importance of a need which has been implied by the customer. For example, if the customer says that his main problem is one of unreliability the salesman might ask an implication question such as 'How much down-time on the line is caused by unreliability?'

The work of Rackham *et al.* has also shown that successful salesmen (at least commercial and industrial ones — the main focus of their research) are more likely to ask 'need-payoff questions' than less successful ones. In this way they shift attention from the problem to the solution, and so obtain a statement of 'explicit need' from the customer. An example of a 'need-payoff question' is 'How much would you save if . . .?' A customer's statement of 'explicit need' might be something like 'I need a way to solve this particular problem — I'm going to get a new supplier'.

Having arrived at customer needs, then some kind of feature/benefit model (e.g. Lidstone 1974) is usually advocated. Rackham, *et al.* are critical of feature/benefit and derivative concepts and terminologies, and give evidence to show that overemphasis on product features and their benefits is often ineffective and sometimes counterproductive. They propose instead a feature/advantage/benefit model. Their work suggests that to be effective a salesman should first get a statement of 'explicit need' before making any product-benefit statements. Such statements should then be related to an explicit customer need. The focus on product benefits which relate to an 'explicit need' is a critical revision to the traditional feature/benefit model.

I shall refer later to other aspects of the work of Rackham, *et al.*, such as the training technology which they have developed and to their findings about effective behaviours at various stages of the sales interview.

The process model

The process model shown in Figure 3.2 is an adaptation of Vroom's (1976) schematic representation of variables used in leadership research. It views sales behaviour as a result of the interaction of three sets of variables: organizational and domestic; personal attributes of the salesman; and customer characteristics and the sale setting.

Figure 3.2   A model of sales effectiveness.

Organizational variables can greatly affect a salesman's behaviour. For example, Poppleton and Lubbock (1977) cite evidence suggesting that organizational structure, recruitment policies, training programmes, supervisory style and the system of payment can affect sales behaviour.

The model indicates that such organizational variables (along with the variables concerning the salesman's domestic situation) interact with the salesman's personal attributes (e.g. personality characteristics, social skills) to determine his sales behaviour.

Domestic variables have also been found to be related to sales effectiveness, and, by implication, to sales behaviour. For example, Holcombe (1922) found that effectiveness was positively correlated with being married and with number of dependants.

The model in Figure 3.2 does not simply refer to sales behaviours reflecting social skills but also to other aspects of sales behaviour such as administration and planning.

Most research on the personal attributes of salesmen as they relate to effectiveness has focused on rather general measures of personality. Such studies have used measures of temperament, interest, intelligence, abilities, motivation and other personality traits cutting across some of these categories (for a review see Poppleton 1975). These personality measures have then typically been related to various criteria of effectiveness (see the criterion problem below), rather than to specific job behaviours or social skills. Such research, aimed primarily at improving sales selection procedures, has been of limited value despite a considerable research effort (e.g. Thayer 1977). More specific differences in behaviour have been found by Chapple and Donald (1947) for sales personnel in different departments of a retail store. The main discriminator between successful and less successful sales personnel appeared to be flexibility and the ability to adjust well to different styles of interaction.

Other personal attributes which can affect sales behaviour are physique and background. For example, physical attractiveness and being liked tend to go together and there is indirect evidence that likeability can be related to sales effectiveness. For example, Mayfield (1972) found that the ability to form close personal relationships was related to a criterion of effectiveness after 1 year for life-assurance agents.

Relevant background factors include any aspect of a person's experience (e.g. education, work history) which might influence a customer. Their effect depends upon the characteristics of the customer. For example, it has generally been found that people like others who are similar to themselves in various ways, including background and appearance.

Clearly, the behaviour of the customer can also have a very large effect on the salesman's behaviour and detailed discussion of the salesman–customer interaction will follow. The social skills of selling are centred around the salesman's ability to cope effectively with customers' behaviour, e.g. by helping the customer to elaborate his needs or by persuading him that his objections to buying the product are without foundation.

A sales interaction takes place in a particular physical and social setting. Any setting has associated rules and conventions which are culture-bound. If rules or conventions are transgressed an unfavourable perception of the other party usually results. Evidence supporting this statement for sales encounters comes from various studies

e.g. Woodside and Davenport 1974; Busch and Wilson 1976; Evans
1964 and Tosi 1966).

The criterion problem

Figure 3.2 clearly shows the important distinction between sales
behaviour and sales effectiveness as defined by the organization. The
part of the model considered so far might be called a descriptive
model, in that its concern is with what variables determine the salesman's behaviour in any particular situation. The other part might be
termed the normative aspect of the model, with organizational
effectiveness being the main concern, and therefore with how salesmen *should* behave if they are to be effective. Investigation of this part
of the model requires a satisfactory measure (or measures) of
effectiveness and this requirement constitutes the criterion problem.

Whether a particular sales encounter has been a successful one for
the salesman and his organization might, at face value, be thought of
as being rather obvious: if a sale results the encounter has been
successful; if not it has been unsuccessful. This simple analysis does
not take account of the following possibilities: (1) the salesman might
have been able to sell more of his product or a more profitable
product line; (2) in the longer term the sale may result in customer
disaffection either because of the manner of the salesman or because
of the customer's later discovery that the product does not meet his
needs. For example, our interviews with sales managers of life-
assurance agents showed that some agents had very high lapse rates on
policies which they had recently sold. The most common explanation
given to us for this phenomenon was that such salesmen often used
techniques which pressurized the customer into buying a policy but
left him feeling uneasy to such an extent that he subsequently
cancelled it. Hence an apparently successful sale can lose a company
money as one or two premium payments do not cover the
administrative costs of issuing a policy. Furthermore, such sales may
well harm the image of the salesman, his company and his product,
and hence affect the likelihood of future sales.

A sale of the type discussed above may, because of a commission
payment system, be judged by the salesman to be a successful one for
him, although it would be regarded as unsuccessful for the company.
The concern of the model is with criteria of success as defined by the
company, but it will clearly be an unsatisfactory state of affairs for

the company if its criterion of success differs from that of the salesman.

Such provisos apart, there are two ways of investigating selling effectiveness. One is to categorize each sales interaction along a continuum of success. Interactions of varying degrees of success can then be examined in terms of those social skills exhibited with a view to highlighting effective patterns of social skill. Alternatively, we can focus on the salesman by distinguishing between successful and unsuccessful ones. Behaviours which typically distinguish between these two groups should then disclose the critical skills of selling. This approach, which is the one I have chosen, requires that there are valid criteria.

The first choice is whether to select a global criterion of effectiveness or to use several different criteria of success (multiple criteria). Examples of global criteria might be overall ratings of effectiveness by the supervisor or the average number of products sold monthly. The latter measure, however, may not give satisfactory comparisons between salesmen with different lengths of service. For example, salesmen may on average gradually increase their number of monthly sales over the first few years. If this is the case then some appropriate statistical adjustment must be made. Another global criterion which is sometimes used is the duration of employment. This makes sense if job tenure is contingent on effective performance.

Global criteria are useful for several purposes, but to help us understand the specific effects of sales behaviours, a number of different, more specific criteria must be used. If required these criteria can be combined to form a single, global composite criterion. For example, Peterson (1953) developed a simple composite criterion which has since been widely used in research investigating the validity of selection instruments. This combines a measure of job tenure with a measure of sales production. Those salesmen who survive for at least a year and who produce more than the median for survivors are classed as successful. Similarly, three other criterion groups may be defined in terms of the two criterion elements.

The different kinds of criteria which have been used in sales research include various kinds of managerial ratings, earnings (where commission or bonus is an element), sales activity measures (e.g. sales volume, sales volume against target and call frequency), job tenure and behavioural analysis data. Appraisal forms can also be useful sources of performance ratings data. For example, Lamont and

Lundstrom (1977) used a sixty-three item appraisal form over a 2-year period to obtain managerial ratings on four different aspects of job performance for salesmen of industrial building materials; technical competence, salesmanship skills, territory management and supportive and developmental strengths.

Ratings on such global categories as salesmanship skills can be broken down to cover more specific areas of job performance. An effective way of doing this is via the critical incident technique (Flanagan 1954). This technique has been used for identifying the critical factors in successful salesmanship by Kirchner and Dunnette (1957) for electrical-equipment salesmen and for life-assurance salesmen in our research programme. Kirchner and Dunnette suggest that this technique can help us to define in objective form the actual behaviour that characterizes successful salesmen. Critical incidents may be defined as 'occurrences that have proved to be the key to effective performance on the job' (Kirchner and Dunnette 1957). They are recorded in the form of stories or anecdotes about how a person handled certain situations. This may be done either on some sort of critical-incident record form, from structured interviews, or with a combination of the two methods. The resulting dimensions may be used as multiple criteria. Studies giving rise to such dimensions are reported in the next section.

The critical skills of selling

Kirchner and Dunnette reported 135 different critical incidents, whilst our studies on life-assurance salesmen have shown almost 200. Each incident may reveal either effective or ineffective performance. Although each incident refers to a highly specific sales behaviour, many of the incidents which were reported were found to be 'over-lapping (and) . . . could be grouped into broader and more meaningful categories' (Kirchner and Dunnette 1957). They found fifteen different categories of critical functions or factors in their salesmen: following up complaints, requests, orders and leads; planning ahead; communicating all necessary information to sales managers; communicating truthful information to managers and customers; carrying out promises; persisting on tough accounts; pointing out uses for other company products besides the salesman's own line; using new sales techniques and methods; preventing price-cutting by dealers and customers; initiating new selling ideas; knowing customer requirements; defending company policies; calling on all accounts; helping customers with equipment and displays; and finally, showing a non-passive attitude.

It is interesting to compare these categories with the eleven which were arrived at as a result of our work. These are:

(1) Showing and generating enthusiasm for the product (e.g. b telling the customer stories or anecdotes from the salesman experience which show the value attached to the product by th salesman).

(2) Paying attention to the customer's needs, feelings and require ments.

(3) Showing integrity and professionalism (e.g. by giving accurat information, carrying out promises, defending company policie and generally showing a belief in the job).

(4) Planning and organizing (e.g. keeping up-to-date records in a organized filing system and planning work in advance.

(5) Persuading and overcoming objections in the sales interview b positive, fluent verbal responses.

(6) Working steadily with relatively little supervision.

(7) Responding positively to the incentives of a commission-only pay ment system.

(8) Devoting time and energy to work activities over and above th minimum requirements (e.g. seeing *all* customers regularly persisting with difficult customers and finding out abou products and selling techniques).

(9) Making effective contacts with prospective customers (e.g. b joining organizations, visiting people who might be sources o clients, helping people in various ways and making onese acceptable to others).

(10) Coping with 'rejection' (i.e. by minimizing its occurrence and b attempting to overcome it).

(11) Showing mood control (i.e. not showing excessive elation when . sale has been made or excessive disappointment when objection are raised).

These results come from two studies conducted in two similar direct-selling life-assurance companies. The results were essentially similar for the two organizations, as was the research methodology. For illustrative purposes this will be briefly described for one of the organizations.

Information was collected by using semi-structured interviews conducted on a one-to-one basis, using the critical incident technique (Flanagan 1954); for a description of the organization, see Poppleton and Lubbock (1977). Interviews lasted on average for about an hour and were carried out by two researchers, each of whom conducted half of the interviews. These were conducted with two area managers, four branch managers, three assistant managers, four unit managers, six senior staff with sales-force responsibilities, fourteen representatives classified as successful, seven as average and fourteen as unsuccessful according to a sales-volume criterion. Critical incidents were tabulated individually, there being 198 different ones. These then gave rise to the eleven broader categories, which represent the effective skills of selling for direct-selling life-assurance salesmen.

As in Flanagan's (1954) study and in that of Kirchner and Dunnette (1957), the broader categories were arrived at intuitively by the researchers on the basis of their 'implicit personality theories' (see Vernon 1964). However, we then attempted to test the validity of these implicit theories by reference to experts, i.e. to a group of sales managers with considerable knowledge of the sales behaviour of a wide range of salesmen. They were asked to judge whether they believed that the groups of critical incident behaviours which we had grouped together were intercorrelated so as to represent a coherent cluster of behaviour to them. There was a high degree of agreement between these experts and the researchers on the nature and composition of the resulting broad categories. Currently we are carrying out similar analyses by questionnaire rather than interview. The data are being analysed via factor analytic techniques, and the analysis is not yet complete. Hence it is not yet possible to say whether the categories showed in our earlier studies will be replicated in the subsequent factor analytic ones. Our categories appear to be somewhat more general than those arrived at by Kirchner and Dunnette. The degree of similarity is clear but there are significant differences, as might be expected for two different sales jobs.

Such categories, which are defined in terms of behaviours, can then be used as the basis for ratings of job performance. For such ratings to be valid they must be clearly defined and the rater must have evidence on each of the specific kinds of job behaviour available; i.e. he must know in considerable detail how a salesman reacts to each of the critical job areas, which implies close observation of his day-to-day behaviour.

I have mentioned earlier that multiple criteria may be combined statistically to give a global composite criterion. However, jobs may have a number of different aspects which bear very little relation to each other. Hence, social skills relevant to one aspect of performance (e.g. making effective contacts) may be different from those used in other aspects, such as persuading others. If the effects of specific areas of social skills training are to be assessed then they should ideally be related to a specific relevant criterion rather than a global composite measure of effectiveness.

Most commonly, however, research studies have used general, rather than specific, criteria. For example, in our initial validation studies on various selection instruments we used average monthly commission earnings as the criterion of effectiveness. Similarly, Baehr and Williams (1968), studying life-assurance salesmen, obtained an overall performance rating measure obtained via the technique of paired comparison as well as measures of mean sales volume, maximum sales volume and job tenure. However, they found that intercorrelations between these criteria ranged from low-negative to high-positive values, i.e. each criterion was measuring an almost unique aspect of effectiveness.

In addition to the need for multiple criteria, different sales jobs require different criteria, because the situations they cause are different in important respects. This is illustrated by comparing Kirchner and Dunnette's (1957) study with our own. The next section is concerned with such differences between the various types of selling.

Types of sales situation and their requirements

Unfortunately, evidence on the critical job dimensions of different kinds of sales jobs is as yet limited. Were this not so, a basis for the classification of sales jobs in terms of their critical dimensions would be available. Because of differences in job requirements in different organizations marketing similar products, such classifications would cut across product lines. For example, the demands on the life-assurance salesman who deals with the general public are widely different from those of the one who deals with insurance brokers.

Because of the paucity of published evidence on these differences in demands, distinctions have been made on theoretical grounds. For example, Lidstone (1974) distinguished between nine categories of sales position. This categorization includes two distinctions which have traditionally been made in the sales literature (e.g. Guion 1965): (1) creative versus non-creative selling; and (2) selling tangible versus intangible products. The former distinction concerns whether or not the salesman must 'create' his own market by finding customers. The latter distinction concerns the nature of the product, i.e. whether it is a consumer durable, a thing which can be seen and used as opposed to something of an abstract nature like life-assurance (an intangible). However, there is virtually no published evidence on the required behaviours for the effective non-creative selling of tangibles (retailing) as opposed to the creative selling of intangibles or the other categories of selling.

Hence, although this classification is used to make statements about the relative difficulty of each type of selling or about the desirable characteristics of each type of salesman, such statements are rarely supported by research studies.

Consequently, much of what follows in this section does not rest on 'hard' empirical data and must be speculative in nature. The remainder of this section will therefore comprise: a rationale for categorizing sales situations; and a model attempting to relate selling technique, effectiveness and the situation. This approach is adapted from Fiedler's (1967) model of leadership effectiveness.

The types of sales situation

Because of differences between selling jobs due to differences in organization structure, nature of the products sold, type of customers dealt with, etc. it is not very useful to distinguish between sales situations in terms of the traditional categorizations of sales job (e.g. Lidstone 1974). Further, any categorization of the selling situation must take account of the fact that each selling interaction can represent a widely different situation in terms of the requirements it places on selling technique, even for the same salesman selling the same product. What then, are the most important specific situational factors?

First, and of greatest importance, is the *acceptability* of the salesman to the potential customer. One of the folklores of selling is that a salesman has to sell himself before he can expect to sell his product. Acceptability is, of course, a function of what the salesman *does and represents* in any interaction, but in relation to the customer's expectations and beliefs about the salesman. There is evidence that salesmen are generally perceived in negative terms (e.g. Evans 1964). However, Evans (1964) found that successful life-assurance agents fulfilled the customer's expectations of insurance expertness, shared similarities in outlook and personal situation with the customer and displayed friendliness towards and personal interest in the customer. Similar results were found by Tosi (1966) for wholesale drug salesmen. Additional supportive evidence comes from Riordan, *et al.* (1977), who found that role congruence (the absolute difference in perceptions of the actual and ideal life-assurance agent) was the most powerful discriminator between sold and unsold customers.

How the initial contact is made is very important for determining acceptability. Research suggests that salesmen are more acceptable to a client if the client makes the initial approach (Poppleton, unpublished observations). There is also evidence that when clients are introduced to the salesman through a third party who is respected by the client, the salesman is more acceptable (ibid.). Similarly, successful life-assurance salesmen were more likely to contact prospective customers through referred leads (i.e. they obtained names of acquaintances and friends of prospective customers from various sources, often from people who were themselves clients). They would then use the name of the 'referrer' in their introduction, so making acceptability more likely. The implication here is that an important sales skill in creative selling is selecting prospective customers who are likely to find the salesman acceptable.

The second important situational variable might be termed the 'power' of the salesman, although it may also be viewed as one aspect of acceptability. Busch and Wilson (1976) see the power of a salesman as being determined by his referent value and his 'expertness'. The powerful salesman has high referent value and a high degree of expertise. Power may also be viewed as a function of the extent to which the salesman has a product or service which is a highly saleable commodity (e.g. because it is the only one of its kind available, it is clearly the cheapest of comparable products, it is clearly superior to other products or it is in short supply). A product which is highly saleable will usually make it easier for the salesman to arrange sales and will put at a premium such selling skills as minimizing interaction time and negotiating.

The third important situational variable is task complexity. At one end of the continuum is simply taking an order for a low-value, echnically simple product (i.e. one which the customer finds easy to understand); while at the other are salesmen who sell complex products or services (often as part of a sales team) to groups of people such as committees or boards of directors. These groups expect that he salesman will spend a considerable amount of time explaining what the product or service can do, that he will have a considerable amount of specialist knowledge and that he should have certain types of specialist help at his disposal.

Each of these three situational variables can be seen as affecting the favourability of the situation for the salesman. The most favourable situation is one in which the salesman is acceptable to the prospective customer, the salesman has high referent value, the product or service being offered is highly competitive and the product or service is easy for the customer to understand. The least favourable situation includes low acceptance, low competitiveness and referent value, and complexity of the product. An intermediate degree of favourableness may also occur.

This contingency model holds that, under clearly favourable or unfavourable conditions to the salesman, the most effective approach will be the 'hard sell'. Under conditions of moderate favourability, the 'soft sell' will be most effective.

A contingency model of selling effectiveness

The distinction between the 'hard' and 'soft' selling approaches is one which has been made several times in the sales literature. For example, evidence for the distinction comes from a study by Miner (1962) on salesmen in the motor-vehicle trade and from our own research on life-assurance salesmen. Table 3.1 shows the chief differences.

Evidence for such a contingency model comes from our own research. One of the most striking findings was that, even within the same company, the most successful salesmen tended to be described as behaving in one of two very different ways. Closer analysis showed that they usually showed one dominant strategy corresponding to either the 'hard' or 'soft' selling approaches described in Table 3.1 and that the predominant strategy was usually associated with the favourability level of the situation predicted above. For example, the highest producer in one life-assurance company we studied used a 'hard' selling approach. He used a standard format, concentrated on

Table 3.1 The hard-selling and soft-selling approaches.

Salesman in:	
Hard sell	Soft sell
Is not particularly concerned with with being liked or gaining sympathy.	Tries to be acceptable to the prospective customer, uses ingratiation techniques.
Comes to the point quickly and makes it clear that time is limited.	Does not hurry the client; gives the impression that he has got plenty of time.
Usually adopts a highly structured approach. May have decided what to sell before meeting the prospective customer.	Shows considerable flexibility, depending on customer's responses.
Focuses on the product and its benefits early in the interview.	Concern in the early part of the interview is with the customer's needs rather than the product.
Uses closing techniques immediately acceptance of the sales idea is shown.	Attempts to get customer to ask for the product rather than using closing techniques.
Dismisses objections quickly or ignores them.	Deals with objections sympathetically and may agree with them, at least in part.
Argues with the customer, openly attempts to persuade him and attempts to dominate the customer.	Gives the customer the impression that he is dominating the salesman.

one type of product and limited calls to a maximum of half an hour. He used a referred lead system, which meant that he was fairly high on acceptability (at least initially), sold to a clientele who were not generally aware of what product competition was available (and did not know how to get access to that kind of information) and made the presentation easy to understand by limiting it to one basic product, which he explained simply and with little detail — i.e. the situation was generally favourable and therefore, according to the model, compatible with a 'hard' selling approach.

The contingency model is, as yet, very much in its hypothetical stage, and is in need of testing in a variety of sales jobs and settings.

The salesman–buyer interaction

In creative selling the salesman approaches many people who, at least initially, do not wish to buy his product. Hence the risk of his sales attempt being rejected is typically high. We found that fear of such rejection was the most common and strongly expressed dislike of the job in life-assurance salesmen. A persistent, high level of rejection usually resulted in high anxiety, which often led to various kinds of job withdrawal behaviours and ultimately to failure (Poppleton and Riley 1977). One salesman, for example, told us that over a 3-year period he had performed reasonably well. However, his rejection rate was very high (nine out of ten approaches resulted in no sale) and this meant that he had to work very hard and that the emotional stress of coping with rejection was very high. He likened the stress to battle fatigue in that the effects of the stress were cumulative. After 3 years he resigned as he felt that he could no longer cope.

Clearly, a key to success in selling over the longer term is to find ways of minimizing the effects of such rejection; one way of doing this is to find prospective customers who are less likely to reject the salesman. This uses two kinds of social skill: initially selecting customers who are most likely to find the salesman acceptable, and using effective referred-lead techniques.

There is a considerable amount of evidence (e.g. Vernon 1964) that people tend to like others who are similar to themselves in various ways. We found ample evidence that successful life-assurance salesmen were particularly adept at selecting people similar to themselves as prospective customers.

A referred lead is the name of a person who might be interested in the salesman's product, given to a salesman by someone. We found that successful salesmen were more likely than unsuccessful ones to consistently use a referred-lead technique to find prospective customers. The success of the technique was found to depend on a number of factors (Poppleton 1974): the source of referred leads; introducing the idea of referred leads throughout sales interviews; capitalizing on opportunities arising during sales interviews; the way in which referred leads were requested; adapting to the stage of the salesman (customer relationship); qualifying referred leads; and the manner of using a referred lead in a subsequent interview.

Successful life-assurance salesmen were particularly likely to consistently obtain referred leads from two sources, namely from other professional men (estate agents, building-society managers and solicitors) and from clients to whom they had sold life assurance. The link with other professional men was almost invariably on the basis that each party put business the other's way whenever possible. In their sales interviews, successful representatives more often said that they introduced the idea of referred leads throughout the interview. They often introduced the idea at the beginning of a sales interview by some such phrase as 'I normally do business by referred leads', thus arousing the expectation in the customer that he will be asked for referred leads. They also paid particular attention to any references to other people which the prospective customer might make during conversation and were more likely to follow up with questions about such people. They were more adept at guiding the prospective customer's conversation towards his friends, colleagues, clubs, associations, etc.

The success rate in getting referred leads was reported to us as varying from as low as 5 per cent in one individual to approaching 100 per cent for others. Technique in asking for leads was certainly partly responsible for such differences. Typically, successful representatives were more likely to: ask for referred leads only when good rapport had been established (usually at the end of the sales interview); ask questions in such a way that the prospect's consent was implied, e.g. 'Perhaps your relations and friends would be interested . . .'; and ask a number of specific questions rather than 'Who do you know who might be interested in life assurance?' Examples of such specific questions were: 'Who do you know who is successful in business?'; 'Who is next in line for your job?' and 'Do you know anyone who is about to buy a house?', etc.

Successful salesmen seemed to show more sensitivity to the dynamic nature of the salesman—customer relationship over time, which affected the way in which they asked for referred leads. Hence, they would be formal in their mode of questioning of new prospective customers but often very informal with established customers. Less successful salesmen were reported more often as being too formal or informal for the situation. Successful salesmen were also more likely to ask a customer for referred leads on *every* occasion he was seen so that an expectation was built up in the customer. Some salesmen reported that they were consequently greeted with customers giving them unsolicited referred leads.

I have mentioned that successful salesmen are more likely to ask specific questions to get referred leads. They are also more likely to qualify (i.e. ask a number of follow-up questions) each referred lead by asking for such information as age, marital status, number of dependants, health, occupation, etc. They are thus able to assess the 'quality' of a referred lead and are better prepared for a subsequent sales interview.

Finally, effective technique included asking the referrer to contact the person whose name they had given to say that the salesman would be contacting them, so making the salesman's initial approach less likely to be rejected.

Along with Rackham, *et al.* we found no evidence supporting the effectiveness of standard, 'trained' openings. Rather, effective salesmen developed their own idiosyncratic, though consistent, ways of starting off interviews, which suited their personality and made them feel most at ease.

The crucial skills, then, are of asking questions (less successful salesmen were reported as taking longer to reach the point of asking the prospective customer relevant questions), listening, and asking further questions. Our findings generally agreed with those selling models emphasizing the importance of getting the prospective customer to talk about his needs before asking him to buy a product. Rackham, *et al.* similarly found that it was more effective for salesmen to hold back and ask implication questions about a customer's statement of a problem, rather than going into product benefits too early. Their research indicates that successful salesmen are more likely to get a customer to state an explicit need before shifting attention from the problem to the solution. This shift is carried out by asking need-payoff questions such as: 'How much money could you save if . . .?'

We found that successful salesmen were more likely to judge the level and amount of information required by the customer correctly. They were less likely to go into technical details that the customer would be unlikely to understand, and generally gave only the amount of detail asked for. This is in line with the findings of Rackham *et al.* that, as salesmen become more experienced (and effective) they focus increasingly on product advantages rather than benefits.

A striking feature of our interviews with successful salesmen was how often they said that they used stories and anecdotes to illustrate the benefits and consequences of having or not having the product. We were given accounts of such anecdotes and they were often very powerful in their effects — even upon research interviewers! Such stories were told with an air of conviction and often used experiences of the salesman himself or of his relatives and friends. Hence the ability to tell stories which move and enthuse another seems to be an important social skill, at least in the groups of salesmen we have studied.

A critical social skill in selling is the handling of objections, i.e. the prospective customer's stated reasons for not wanting to buy the product. Successful salesmen more often reported that they listened to objections without interruption, and then showed the prospective customer that they understood them and took them seriously. Less successful salesmen were more likely to attempt to rebut an objection immediately, sometimes without hearing it out in full. Further, more-successful salesmen were less likely to get into a lengthy argument on any particular objection. As one successful agent said, 'I give the minimum information necessary to counter any objection'. Another finding was that successful representatives were more likely to use (and to have at their disposal) stock phrases for a particular kind of objection.

Although successful representatives were generally more likely to allow the prospective customer to air his objections in full and to listen with understanding, another type of response of some unsuccessful representatives resembled this. However, this group typically reported feeling such understanding and sympathy for the prospective customer's point of view (i.e. that his objections were valid) that they did not attempt to overcome the objections raised.

Another social skill of considerable importance in the selling interaction is 'the close' (i.e. the achievement of the salesman's objective – usually a completed sale). 'Closing techniques' typically refer to ways of asking the prospective customer to commit himself to a definite purchase. Such techniques include the direct request (e.g. 'Will you sign the contract here?'), the 'alternative close' (e.g. 'Would you like a "with profits" or "without profits" policy?') and the assumptive close (e.g. 'You have clearly decided to buy this product, but we've got to decide on a delivery date, haven't we?'). Closing techniques are typically part of most current training programmes. However, their effectiveness is in some doubt when the scanty evidence on their efficacy is reviewed.

This is shown most clearly in about half a dozen studies carried out by Rackham, *et al.* (unpublished observations), who found that when the product being sold was expensive then the use of closing techniques was *negatively* related to the likelihood of making a sale. However, they were positively related (though only slightly) to effectiveness in a group of retail salesmen of low-cost electronic equipment. For this latter group, up to two closing attempts per sales interview were positively related to the likelihood of a sale being made.

Perhaps the clearest finding to emerge from research in this area is that the particular techniques used at each stage in the sales interview are typically adapted by the successful salesman to suit his personality. Those salesmen who perform in the correct training-school manner without being sufficiently 'true to themselves' are unlikely to succeed. Thus sales interviews, particularly when conducted by successful salesmen, tend to fit neatly with the traditional view that a sale should follow a definite sequence of steps. This traditional view holds that every sales interview should have a sequence such as the following (Lidstone 1974): opening, creating a good impression, gaining the customer's attention, exploring the customer's needs, presentation, benefits, objections handling and the close. The successful salesman was more likely to show flexibility in responding to the customer and might omit one or more stages and show variations in the order in which stages are tackled. Similarly, Argyle and Lydall (in Argyle 1972) found no evidence for the traditional theory of sequence of events at a sale for retail saleswomen.

At a very general level I have distinguished between the hard and soft selling styles. Little work has been done on the relationship between personality and sales behaviour or style. However, our impression from our own research is that successful salesmen using the 'hard sell' approach tend to be low in their need for affection and acceptance from others, high in dominance (particularly in its aspects of competitiveness, ambition and a tendency to aggression) and emotionally resilient (and hence capable of coping with rejection), but not particularly high in such social skills as empathy. Conversely, the successful 'soft sell' salesman is typically high in the social skills of selling (empathy, showing sensitivity towards others, sympathetic listening, etc.) and has a high need for the acceptance of others.

There is also evidence that client characteristics help to determine what selling behaviours will be effective. Specifically, when the prospect's product knowledge is low and when his educational and intellectual levels are low then the 'hard sell' approach is more likely to be effective. Conversely, with educated and informed prospects, the 'soft sell' approach is likely to be most effective. These contentions are based on our evidence that the 'package' selling of life assurance ('hard sell') is most effective in selling to low-income and low-educational level groups. This also receives indirect support from the findings of Rackham, *et al.* on closing techniques if one assumes that high-value products are more likely to be sold to sophisticated buyers.

Much of the research cited so far has concerned creative selling, and the selling of life assurance in particular. Retail selling, however, has also received the attention of researchers and does raise a number of interesting issues. Particularly significant is the importance of non-verbal cues. For example, a customer may signal by means of eye-gaze behaviour that he or she wishes either to make contact or avoid contact with the salesman. Argyle and Lydall (in Argyle 1972) have also found that the way in which the salesman handles goods can convey a message, as illustrated by the fact that in some shops the more expensive goods were handled more reverently. Other NV cues such as dress and bearing are likely to affect the categorization of customers and resulting sales strategies.

According to the contingency model of selling, the retail situation is likely to be a relatively favourable one for the salesman. Typically, the customer chooses the retailer and hence the retail salesman is likely to be acceptable to the customer. Typically, the product sold will be relatively simple in concept, which also adds to the favourable-ness of the situation. Finally, the salesperson may have been chosen in part because of his power (i.e. perceived expertise and referent value). Hence, the retail situation will often represent the most favourable selling situation, and when it does so the model predicts that the most effective retail salesperson will be one using a 'hard selling' style.

Implications for current training practices

Training programmes for sales personnel should stem from an analysis of training needs via job analysis. Such analyses should use relevant techniques for isolating those parts of the job which require training, such as the critical-incident technique. If such analyses are carried out it will be found that most of the critical job requirements use various kinds of social skill and such procedures would give rise to training programmes with the prime focus being on the acquisition and development of social skills.

Our experience suggests that only a minority of sales training programmes have such a focus, and that social skill training often plays a minor role compared with product knowledge (technical) training. This is despite the fact that studies (e.g. Baier and

Dugan 1957) have shown that product knowledge may have little or no relationship to sales effectiveness.

Further, when programmes include a significant element of social skills training with such techniques as demonstration, role playing or on-the-job coaching, they are rarely based on a detailed, systematic analysis of the sales job in question to elicit specific behavioural and skill requirements.

An exception to this statement is the work of Rackham, *et al.* (unpublished observations). They have developed a training technology based on behavioural analysis which has been briefly described earlier in this chapter. Their technology includes the training of sales managers in behavioural analysis techniques so that they are able to recognize various types of sales behaviours, interpret such behavioural data and subsequently give feedback. Behavioural data is analysed with a central computer, resulting in a report on which behaviours work well for a particular salesman, what changes are taking place in his behaviour and other data of use to the trainer. However, this work is very recent and there are no published data available on the evaluation of such training programmes.

Other implications of work on the social skills of selling

Perhaps the most important questions to be asked about the kinds of social skills to which I have referred are (1) how far can they be trained and (2) how far are they relatively ingrained aspects of personality which are relatively resistant to attempts to change them?

As yet, the evidence relating to such questions is rather scant. Certainly, a striking phenomenon which sales trainers encounter is that the behaviours of some people are extremely resistant to change. There is a danger in assuming that a sales training programme emphasizing social skills will automatically have a beneficial effect on performance, even when it is well designed along lines such as those advocated by Rackham, *et al.* I believe that it is important to supplement a social skills training programme with a recruitment policy which pays attention to the levels of social skills present in job applicants. The analysis of the social skills of selling can be used to develop instruments which can be used to measure the levels of these kinds of skills in job applicants. This is a task which we have been working on over the past 5 years to design a sales aptitude test.

Conclusion

Very little detailed research has as yet been carried out on the social skills of selling. Hence, much of what I have written rests heavily on our own research into one type of salesman, and care should be taken not to over-generalize such findings.

The relative paucity of research data in this area has also meant that some of the ideas presented have been speculative; hopefully, it will not be long before there is empirical evidence bearing upon them.

References

Argyle, M. (1972). *The Psychology of Interpersonal Behaviour*. London: Penguin, 2nd edn.

Argyle, M. and Kendon, A. (1967) The experimental analysis of social performance. *In* Berkowitz, L. (ed.). *Advances in Experimental Social Psychology* – volume 3. New York: Academic Press.

Baehr, M. E. E. and Williams, G. B. (1968). Prediction of sales success from factorially determined dimensions of personal background data. *J. Appl. Psychol.* 52, 98–103.

Baier, D. E. and Dugan, R. D. (1957). Factors in sales success. *J. Appl. Psychol.* 41, 37–40.

Bonoma, T. V. and Felder, L. C. (1977). Nonverbal communication in marketing: toward a communicational analysis. *J. Market. Res.* 14, 169–80.

Busch, P. and Wilson, D. T. (1976). An experimental analysis of a salesman's expert and referent bases of social power in the buyer-seller dyad. *J. Market. Res.* 13, 3–11.

Chapple, E. D. and Donald, G. (1947). An evaluation of department store salespeople by the interaction chronograph. *J. Market.* 12, 173–85.

Evans, F. B. (1964). Dyadic interaction in selling – a new approach. *Am. Behav. Sci.* 6, 76–9.

Fiedler, F. E. (1967). *A Theory of Leadership Effectiveness*. New York: McGraw-Hill.

Flanagan, J. C. (1954). The Critical Incident Technique. *Psychol. Bull.* 51, 327–58.

Greenberg, H. and Mayer, D. (1964). A new approach to the scientific selection of successful salesmen. *J. Psychol.* 57, 113–23.

Guion, R. M. (1965). *Personal Testing*. New York: McGraw-Hill. Series in Psychology.

Holcombe, J. M., Jr. (1922). A Case of Sales Research: Report on first steps in a study of selection of life insurance salesmen. *Bull. Taylor Society* 7, 112–21.

Kirchner, W. K. and Dunnette, M. D. (1957). Identifying the critical factors in successful salesmanship. *Personnel* (September–October), 54–9.

Lamont, L. M. and Lundstrom, W. J. (1977). Identifying successful industrial salesmen by personality and personal characteristics. *J. Market. Res.* 14, 517–29.

Lidstone, J. (1974). *The Nature of Salesmanship*. London: Gower Press.

Lombard, G. G. F. (1955). *Behaviour in a Selling Group*. Harvard University Press.

Mayfield, E. C. (1972). Value of peer nominations in predicting life insurance sales performance. *J. Appl. Psychol.* 56, 319–23.

Miner, J. B. (1962). Personality and ability factors in sales performance. *J. Appl. Psychol.* 46, 6–13.

Peterson, D. A. (1953). Review of aptitude index No. 825. *In* Buros, O. K. (ed.). *Fourth Mental Measurements Yearbook*. Highland Park, NJ: Gryphon, p. 822.

Poppleton, S. E. (1975). *Biographical and Personality Characteristics Associated with Success in Life Assurance Salesmen*. M. Phil. thesis, Birkbeck College, University of London.

Poppleton, S. E. and Lubbock, J. (1977). Marketing aspects of life assurance selling. *Eur. J. Market.* 11, 418–31.

Poppleton, S. E. and Allen, E. (1978). *The Construction of a Sales Selection Test*. Presented at the XIXth International Conference of Applied Psychology, Munich.

_____ and _____ (in press). *The Handbook for the Poppleton–Allen Sales Aptitude Test*. Slough: National Foundation for Educational Research.

Poppleton, S. E., Riley, D. and Lubbock, J. (1977). *The Personality Characteristics of Life Assurance Salesmen I*. Given at the Occupational Psychology Annual Conference, University of Sheffield.

Poppleton, S. E., Riley, D. and Allen, E. (1978). *The Personality Characteristics of Life Assurance Salesmen. II*. Given at the Occupational Psychology Annual Conference, University of Cambridge.

Rackham, N. and Morgan, R. G. T. (1977). *Behaviour Analysis in Training*. London: McGraw-Hill.

Riordan, E. A., Oliver, R. L. and Donnelly, J. H., Jr. (1977). The unsold prospect: dyadic and attitudinal determinants. *J. Market. Res.* 14, 530–7.

Thayer, P. W. (1977). Somethings Old, Somethings New. *Personnel Psychol.* 30, 513–24.

Tosi, H. L. (1966). The effects of expectation levels and role consensus on the buyer-seller dyad. *J. Business* 39, 516–29.

Trower, P., Bryant, B. and Argyle, M. (1978). *Social Skills and Mental Health*. London: Methuen.

Vernon, P. E. (1964). *Personality Assessment*. London: Methuen.

Vroom, V. H. (1976). Leadership. *In* Dunnette, M. D. *Handbook of Industrial and Organisational Psychology*. New York: Rand McNally.

Woodside, A. G. and Davenport, J. W., Jr. (1974). The effect of salesman similarity and expertise on consumer purchasing. *J. Market. Res.* 11, 198–202.

Woodward, J. (1960). *The Saleswoman*. London: Pitman.

4 Negotiation and bargaining

IAN MORLEY

Introduction

'Negotiation is an enduring art form. Its essence is artifice, the creation of expedients through the application of human ingenuity.'   (Winham 1977, p. 87.)

Following Morley and Stephenson (1977) we may note that *negotiation* may be defined as a process which has the following characteristics:

(1) Negotiation is a process of joint decision making concerned with the form of action jointly to be taken to define or redefine the terms upon which persons or parties will 'do business'. Negotiation is thus not the same as argument or persuasion or debate. Furthermore, although negotiation may include third parties (as in inquiry, conciliation, or mediation) it is to be distinguished from arbitration (and other forms of judicial settlement) in which the main protagonists are no longer fully responsible for the agreements which are obtained.

(2) 'Conflict' or 'struggle' is an essential component of negotiation since participants place alternative agreements in a different order of value or preference. However, negotiation engages 'mixed-motives', since attempts to achieve personal or party goals are constrained by some degree of common interest. Minimally, the constraint may be to keep the other at the 'bargaining table' (as an exercise in propoganda, gaining intelligence, or whatever) or to reach an agreement: to get something rather than nothing. Or there may be common concern with the 'efficiency' as well as

the 'distributional' aspects of bargaining in recognition that some agreements are better for both parties than others (Walton and McKersie 1965).

(3) Negotiation is thus a form of 'incomplete antagonism' or 'precarious partnership' which allows each participant the opportunity to manipulate perceptions of common interest in an attempt to achieve private goals 'by threatening to destroy some common good or by promising mutual gain' (Snyder and Diesing 1977). More generally, it may be said that negotiators may attempt to exploit asymmetries in the situation generated by unequal resources and by unequal stakes in the issues. Therefore, in a rough way the outcomes of negotiation may be expected to reflect the costs each actor incurs in rejecting rather than accepting a given demand (Morley 1979; Atkinson 1975). Costs of this kind are a major determinant of bargaining 'resolve' (Snyder and Diesing 1977; Lockhart 1979).

(4) Each party may be motivated to misrepresent its position in one way or another in an attempt to outwit its opponent and defend its own interests. In other words, negotiation is a form of strategic decision making in which it may be necessary to assume that: 'rational actors try to project desired images, whether accurate or not, and skeptically view the images projected by others' (Jervis 1970, p. 15). Practitioners cannot escape questions on why they should believe what others say or on what can be inferred from others' behaviour. For this reason alone there is an irreducibly psychological component in the process of negotiation (Morley 1979). The costs of accepting or rejecting a given demand have to be worked out as negotiation proceeds. Process is thus 'practically inseparable from power since it is through process that the true power relations become manifest in the parties' values and perceptions' (Snyder and Diesing 1977, p. 281).

(5) To some extent negotiation includes talking about a relationship before doing anything about it. This may include detailed exposition of a case, followed by long and complex argument; in other cases it may be little more than an exchange of bids. The point is that strength of case provides an additional source of assymetry in negotiation and helps to structure perceptions of resolve (Morley and Stephenson 1977).

Not all negotiation is conducted 'in good faith'; this is commonly observed. For example, it is often stated that the post-war disarmament conference used gamesmanship rather than serious negotiation. The distinction between 'serious' and 'non-serious' kinds of negotiation may be recognized by defining *bargaining* as that form of negotiation in which there is: 'an operative desire to clarify, ameliorate, adjust or settle the dispute or situation.' (Lall 1966, p. 31.) In other words, bargaining is negotiation for agreement (Morley and Stephenson 1977).

To describe negotiation as an exercise in which parties struggle to exploit asymmetries of interest and power, each knowing that the other may disguise or misrepresent their real position, is to recognize that skill in negotiation includes artifice in the sense of strategic manoeuvre, deception, cunning and craft: to recognize that negotiation may be considered 'the management of people through guile' (Winham 1977, p. 87). But the management is limited by 'rules of accommodation' such as:

> 'unambiguous lies must be avoided, explicit promises have to be kept, invective is never to be used, explicit threats must not be issued, agreements in principle must not be blatantly violated when it comes to the execution of details, and mutual understandings must not be deliberately misconstrued later on.'
>
> (Ilké 1964, p. 87.)

The rules may be established by tradition or worked out by the parties themselves. Different rules may govern different kinds of negotiation or negotiations with one person rather than another. For example, it is common for textbooks on industrial relations to stress that a negotiator's effectiveness depends crucially on his reputation for honesty and integrity (Walton and McKersie 1965; Marsh 1974). In contrast, diplomacy has been viewed as an enterprise in which: 'a diplomat's words must have no relation to actions — otherwise what kind of diplomacy is it? . . . Good words are a concealment of bad deeds. Sincere diplomacy is no more possible than dry water or iron wood.' (Stalin, quoted in Jervis 1970, pp. 69–70.) More generally, according to Marsh (1974), 'It is not expected that the negotiator will always tell the truth or tell all of the truth' (p. 193).

Negotiation has some of the characteristics of a 'game of strategy'. But it is a 'game' characterized by various kinds of uncertainty (Winham 1977; Morley, in press). Possible moves are not always specified in advance of the game. Bargaining power is highly situation pecific (Batstone, *et al.* 1977; Lockhart 1979) and it may require considerable 'resourcefulness' to construct moves which apply pressure in any particular case (Lockhart 1979). Many negotiations are also extremely complex in terms of goals, constraints, technical details, number of parties and the like. Negotiators may not have time to study important documents and work out what is going on. There may be major difficulties in structuring decisions so that clear statements of preference can be made: for example, see Douglas's (1962) studies of industrial peacemaking and my discussion (in press) of the difficulties which faced negotiators in the Paris Peace Conference of 1919. Negotiation thus includes artifice in a second (and newer) sense: artifice as constructive skill.

As negotiations become more complex, questions of strategy (how to outwit an opponent) may be subordinated to questions of structure (how to develop common perceptions or agree a framework of broad objectives and principles) (Winham 1977; Iklé 1964). According to Winham (1977):

> 'A more appropriate model for negotiation in a complex situation is one that replaces strategy with search for information, and is concerned with process as opposed to outcome. . . . The process of negotiation involves a search for acceptable solutions, where strategy is more a matter of forestalling the consideration of certain unattractive solutions than a matter of extracting a change of position from an adversary.'
> (p. 97.)

One consequence of complexity is that negotiation may begin without much idea of what would constitute an acceptable agreement (Iklé 1964; Warr 1973; Snyder and Diesing 1977; Morley, in press). There is, therefore, ample room for artifice as 'the product of art' and as a distinctively 'human skill'. Negotiation includes argument, although argument of a rather special kind. But to illustrate I shall cite Harrod's (1953) *The Life of John Maynard Keynes*, which includes the rather striking example of a case in which Lloyd George reversed the British position at the Paris Peace Conference without the other participants noticing the changes which were made. Given half a sheet of notepaper (from Keynes) outlining possible arguments, Lloyd George read it quickly:

> 'and proceeded on the same lines as before. . . . But gradually, as they listened, a gentle trickle of thought of a new kind began to

appear. . . . And then, slowly, as he took plenty of time in making his case, the whole trend was transformed, and he was soon using all Keynes' arguments on the opposite side. . . . It was the finest example . . . of cooperation between two master minds to achieve what at first seemed quite impossible.'   (p. 240.

Most 'checklists' of negotiation skills do little more than reflect the central characteristics of the negotiation task, as outlined so far. Typical contributions* include items such as: 'choosing strategies', 'staying in command even when exposed to severe pressure', 'perceiving and exploiting power', 'setting goals', 'diagnosing the nature of the conflict', 'stating problems and objectives clearly', 'evaluating possible solutions constructively', 'looking at issues from the viewpoint of the other side', 'facility in argument' and so on. The reader is often given neither adequate understanding of the difficulties of the task, nor advised how the various jobs are to be done. A worst he is left with the feeling that what is required is: 'a magica combination of personality and organisational skill' (Margerison and Leary 1975, p. 46). At best he is given a general appreciation of the kinds of ingenuity which will have to be explored if more-adequate understanding is to be gained.

What seems to be required is an account which links behavioura advice to a theoretical treatment of the nature of the negotiation task. Here I hope to contribute to the enterprise in three main ways.

(1) The first part of the paper outlines a framework which view negotiation in terms of 'core' processes of information interpretation, influence and decision making. The treatment relie heavily on the work of Snyder and Diesing (1977) and Lockhar (1979), and emphasizes decisions which shape the general form negotiation will take. Concern with 'structure in process' has bee one enduring focus in the social psychology of negotiation (Morley and Stephenson 1977) and attention will be paid to the ways in which central processes follow different paths in differen situations.

(2) The second part of the paper reviews empirical studies demons trating factors conducive to 'success' in negotiation groups. Particular consideration will be given to the questions: d successful negotiations go through stages which unsuccessfu

* Useful treatments are given by Pedler (1977), Margerison and Leary (1975) Atkinson (1975) and Marsh (1974).

negotiations do not? Do successful negotiators do things which unsuccessful negotiators do not?

(3) The third part of the paper summarizes what has been said and places the discussion within a wide social context.

Snyder and Diesing's model of negotiation

Figure 4.1 is a schematic representation of certain aspects of the processes of information interpretation, influence and decision making as they occur in negotiation. There are several information processing tasks, and situational factors (such as complexity) would be expected to limit severely the efficiency with which they are performed. Each of the 'core' processes will now be considered in turn.

Information interpretation

Three aspects of information interpretation are given a central place in Snyder and Diesing's model: definition of the situation; interpretation of feedback; and revision of expectations. In each case it is emphasized that human beings have limited processing power and are forced to focus upon aspects of bargaining defined as important by perceptions of the world in general, and images of the other in particular (George 1972; Jervis 1976; Kinder and Weiss 1978; Lockhart 1979).

Definition of the situation   Conflict begins when some change in existing circumstances (whether deliberate or inadvertent) creates a situation in which one actor (person or party) must confront another. The 'challenge' may come from a coercive act or other 'violations of tolerance', as in crisis bargaining between nations (Snyder and Diesing 1977; Lockhart 1979) and 'hostage negotiations' (Miron and Goldstein 1979). Or the conflict may begin with a negotiation phase (Warr 1973; Morley and Stephenson 1977). But however the conflict begins actors will attempt to plan suitable responses.

Negotiators interpret new information in terms of pre-existing systems of belief ('amplification') and search for explanations of what has occurred. Here I want to emphasize the part played by actors' philosophical* beliefs about the values which are likely to be

---

* The term 'philosophical' is taken from George (1972).

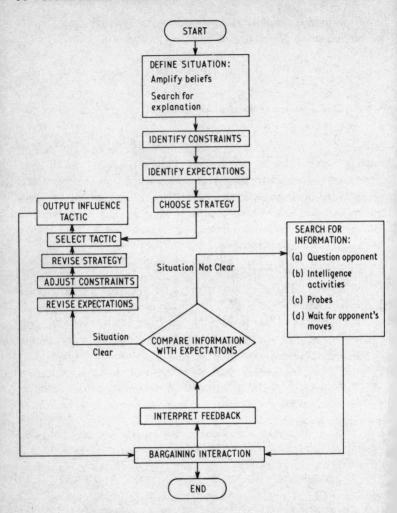

Figure 4.1 Snyder and Diesing's model of negotiation (adapted from Snyder and Diesing 1977, p. 413, Figure 5.3 © 1977 Princeton University Press).

threatened and the opportunities likely to be present in the political system of which they form a part (George 1972; Lockhart 1979; Batstone, *et al.* 1977). In particular, beliefs about the fundamental nature of conflict have far-reaching implications.

For example, Batstone, *et al.* (1977) have reported that staff stewards have less articulated and less radical conceptions of industrial conflict than stewards representing the shop floor. Of the stewards polled:

'almost a quarter of the staff stewards expressed ignorance of any union principles, while almost a third defined them primarily in organizational terms, such as the right to strike. In contrast, many shop-floor stewards referred to improved living standards, ideas of social justice, and the prevention of exploitation. A quarter made specific reference to ideas of socialism and workers' control.'

(Batstone, *et al.* 1977, p. 27.)

In some cases the ideology may be integrated by way of a 'master script' formulated at such a general level that it will explain just about any aspect of an opponent's behaviour (Abelson 1973*). John Foster Dulles' image of the Soviet Union appears to have been controlled by a 'script' of this kind. Consequently:

'If the Soviets made an active move, this proved that they were aggressive; if they were conciliatory, this proved that they were in internal economic difficulty due to the inherent unworkability of Communism; if they yielded on some issue, this proved they were bluffers.' (Snyder and Diesing 1977, p. 331.)

Factors of this kind mean that in crises bargaining between nations, each usually misperceives the opponent's intentions, misinterprets what he has to say, and miscalculates the effects of their own actions (Snyder and Diesing 1977, p. 81). Apparently it is extremely difficult for actors to appreciate that:

'signals so clear to them can be interpreted radically differently by others. . . . Actors often cannot believe that others can fail to see them the way they see themselves and that others may interpret their signals against a background of beliefs holding that the signalling state is deceitful and aggressive.' (Jervis 1970, p. 135.)

Forming expectations of others' responses Estimates of how the other will respond reflect the values that are seen to be at stake. For example, Dean Acheson held the view that the Berlin Crisis of 1961 had nothing to do with West Berlin *per se*. Rather, Khrushchev's

---

* Abelson's 'ideology machine' is a computer program which attempts to show how a right-wing ideologue might respond to questions of the form 'If a Communist nation attacks West Berlin, what should America's NATO allies do?' A less technical account is given in Boden (1977).

intention was to test the general American will to resist. Consequently, there was nothing to negotiate. Acheson argued, therefore, that willingness to negotiate the issue of West Berlin would be taken as a sign of weakness, and make the crisis that much worse.

The example may also be used to make a rather different point, as Snyder and Diesing (1977) have shown. Negotiators often hold instrumental beliefs about ends–means relationships which dictate appropriate political action (George 1972). Broadly speaking, they develop general preferences for accommodative or coercive moves. Further, philosophical and instrumental beliefs combine to produce the systems usually labelled as 'hard-' or 'soft-line' (Snyder and Diesing 1977; Lockhart 1979).

Here, what is important is that hard-line actors — such as Acheson — and soft-line actors — such as Chamberlain — tend to develop images of the opponent which differ in a number of systematic ways. Acheson's analysis of the Berlin Crisis instantiates a master script in which the opponent's moves are seen as part of a general programme of expansion. The immediate objectives may be seen as contributions to that goal or as tests of one's own resolve. Typically, the hard-liner sees his opponent as deliberately 'trying it on' (opponent's options are open) rather than defending central values (opponent's options are closed). It is assumed that any concession will be exploited to the full, but that the opponent will back down when confronted with firmness and resolve.

In contrast, Chamberlain's explanations of Hitler's behaviour illustrates a master script in which both parties have limited and (more or less) legitimate goals. Despite appearances, the opponent is not engaged in an unlimited pursuit of power but willing to compromise, now and in the future. Agreement may, therefore, be one step towards a more general settlement, understanding or détente ('Peace in our time').

Each system of belief implies a different kind of risk. Hard-liners run the risk that opponent's central values *are* at stake and that conflict may increase in magnitude and direction. Agreements may be difficult or impossible, given an inappropriate mix of coercive and accommodative moves. Soft-liners run the risk that the opponent is not, after all, willing to compromise and that concessions thought valuable will be seen as insultingly small (Snyder and Diesing 1977; Lockhart 1979).

Clearly, this account is grossly oversimple. Snyder and Diesing

(1977) and Lockhart (1979) go into much more detail on hard-line and soft-line systems of belief. But what has been shown is that inferences of intention are crucial in negotiation, as in other forms of political life (Jervis 1976).

To elaborate, assessments of intention help to determine answers to questions on whether an issue in dispute can be treated in isolation. How will the other implement an agreement once it has been obtained? (Jervis 1976.) To take the first question first, the negotiator must decide whether the issue should be treated on its merits, so to speak, or as a symptom of a more basic conflict yet to be articulated, and waiting to come to the fore. According to Jervis:

> 'If the underlying conflict is severe and cannot be dissipated by conciliation, then concessions, even if warranted by the facts of the immediate case, may lead the other to raise further demands. . . . If, on the other hand, the actor feels that the basic issues can be ameliorated, he will usually be well advised to try to set relations on a more cooperative footing rather than try to negotiate about the narrow issue that is the immediate subject of contention.'
>
> (Jervis 1976, pp. 44–5.)

Second, many agreements run into difficulties when they are put into effect. There may, of course, be many reasons for failures of this kind, but it is axiomatic that the contract should be clearly understood and spelled out in enough detail. Would the other take advantage of a technicality? Would he accept the spirit of the agreement in areas not explicitly put into words? The point is:

> 'negotiators need to assess their opponents to try to see not only how much they can get out of them but also how much they *need* to get out of them. The greater their fear and mistrust, the more detailed they will want the contract to be and the more guarantees they will believe they must exact.'   (Jervis 1976, p. 45.)

Information search and the revision of expectations   Negotiation is an example of decision making under uncertainty in which values, interests and power relations have to be worked out as arguments are exchanged and moves are made. The uncertainty may be increased by complexity of the kind described by Winham (1977) and Morley (in press); or deliberately cultivated in an attempt to outwit an opponent. The concept of negotiation as artifice includes attempts to impose order and certainty and to work out what is going on.

However, recent work in psychology and political science shows that, in a sense, decision makers exercise *too much ingenuity* in sorting what is essential from what is not (Steinbruner 1974; Jervis 1976; Kinder and Weiss 1978). Actors form strong categorical judgements despite the uncertain probabilistic nature of the environment they face (Steinbruner 1974). New information is assimilated to existing beliefs so that actors see what they expect to see. Evidence to confirm the image is quickly found and disconfirming evidence denied, distorted or ignored. That is, actors show 'premature cognitive closure' (Jervis 1976).

Some people require overwhelming evidence before they will change their views (Holsti 1967); others are more open-minded. Snyder and Diesing (1977) mark the difference by saying that at some point the search for constancy becomes 'irrational'. Jervis (1976) prefers the term 'excessive cognitive closure'. But whatever the preferred term the shift occurs 'as persistence and denial come to dominate openness and flexibility' (Kinder and Weiss 1978, p. 710).

Snyder and Diesing (1977) develop the distinction in terms of rational and irrational 'bargaining modules', which describe 'rather closely' the behaviour exhibited in crisis bargaining. But in my view the 'modules' are of some general importance (Morley, in press). Figure 4.1. shows the rational module (in slightly amended form).

The point to note is that once bargaining is under way, interpretation of new messages:

'is based on the developing pattern of information itself. It is based, not on what we "know" the opponent's fundamental characteristics and ultimate aims are, but on the specific pattern of his overt statements and actions *this time*. New information is, as always, interpreted by its fit with what we "know", but what we know comes from the bargaining process rather than from the image.'

(Snyder and Diesing 1977, p. 335; my emphasis.)

Thus the boxes labelled 'interpret feedback' and 'revise expectations' are not directly linked with the box labelled 'define situation; amplify beliefs; 'search for explanation'.

Contrastingly, the irrational module shows what happens when: 'the bargainer "knows" what is going on and is not going to be fooled by any new information' (Snyder and Diesing 1977, p. 337). Essentially, strategy is not revised in an irrational bargaining module since there is only one possible strategy to use. Expectations follow directly

om the initial system of beliefs. The box 'interpret feedback' is
nked to the box 'define situation' so that all new information is made
onstant with the initial image.

Snyder and Diesing (1977) have provided an extremely useful dis-
ussion of the key characteristics of bargaining which is rational in
his kind of information-processing sense. The essence of the process is
rtifice in the sense of a problem-oriented search:

1) The rational bargainer has low confidence in his initial definition
of what might be going on.

2) Consequently, he initiates an active search for information. In
this context Snyder and Diesing cite with approval Ulbricht's use
of coercive probes (by which they mean 'limited moves against
specific targets that are withdrawn if they meet serious
resistance') in the Berlin Crisis of 1958–62. Presumably, the
bargainer will also search for linguistic cues of the sort outlined
by Walton and McKersie (1965, pp. 96–7). (For example,
Walton and McKersie suggest that the statement 'this is our final
offer' is stronger than the statement 'we must present this as our
final offer'. There seems little doubt that negotiators do use
cues of this kind, but Walton and McKersie's examples should
be treated as illustrating possibilities rather than providing
definitive guides to behaviour.) Lockhart (1979) refers to the
activity as 'goal decomposition' since the object of the exercise is
to determine which aspects of the other's position are flexible,
and by how much. Non-verbal cues, such as increases in the pace
or loudness of speech, increased restlessness, clenched fists, etc.
may also provide useful information (Miron and Goldstein 1979;
Walton and McKersie 1975).

3) Wishful thinking is produced by questions such as 'Is my strategy
working?' Accordingly, the rational bargainer concentrates his
effort in an attempt to work out the strategy of the opponent. For
example, Snyder and Diesing argue that in 1914 Germany had
plenty of information to suggest that Britain would fight. But the
Germans did not ask what British strategy was and misperceived
'the negative information that forced its way through their
filters'.

4) 'Background' characteristics of the image of the other are
resistant to change. Adjustment of expectations is thus a 'limited
affair': 'What is adjusted is the estimate of the opponent's specific

aims, interests, degree of resolve and capabilities in a particular conflict' (Snyder and Diesing 1977, p. 336).

Snyder and Diesing's irrational bargainer is irrational in two kinds of way. First, he demonstrates excessive cognitive closure. He holds a rigid system of beliefs and his beliefs 'dominate' his behaviour. Second, his beliefs are organized so that all considerations point to the same strategic choice. It is this latter characteristic which Jervis regards as 'irrational' (Steinbruner would say 'non-analytic') since the real world is not so benign (Jervis 1976, p. 130). If 'nothing is to be sacrificed' we can be fairly sure that we are looking at a case of 'belief system overkill' which minimizes the conflict between different kinds of constraint.

This kind of conflict is at the heart of Walton and McKersie's (1965) *Behavioral Theory of Labor Negotiations*. Negotiators who are 'nearsighted' or 'myopic' tend to set up objectives and see only one way of getting there. They do not adequately assess the limits within which business may take place. A further consequence is that opportunities are missed to 'upgrade common interests' and achieve long term goals. Bargainers who strive for irrational consistency, in Jervis' sense, are unlikely to engage in what Walton and McKersie (1965) call 'integrative bargaining'. Essentially, they are unlikely to proceed by 'trying out new combinations of ideas in an effort to move the negotiation along' (Winham 1977, p. 101).

Some idea of the nature of the task is given by Walton and McKersie's account of a management response to a demand for a guaranteed annual wage. Instead of engaging in pressure bargaining to preserve the flexibility of managerial response, management:

(1) Diagnosed union concern about the inadequacy of unemployment compensation.
(2) Realized that new laws would be passed unless changes were made.
(3) Anticipated that new laws would mean new obligations for the company.
(4) Initiated moves which led to payment of supplementary employment benefits, a substitute acceptable to both sides (Walton and McKersie 1965, p. 156; Jervis 1976, pp. 46–7).

The problem-solving activity thus contained two elements: diagnosis

f the basic cause for the union demand and an active search for a
trategy that satisfied defined goals or constraints. Miron and
Goldstein (1979) also stress the importance of diagnostic skills in
egotiation including hostages; e.g. make sure there is a hostage, go
hrough a systematic checklist of information about the perpetrator.

In many cases, particularly multilateral negotiations, negotiation
nay begin without much idea of what would constitute an acceptable
greement (Iklé 1964; Winham 1977; Snyder and Diesing 1977; Warr
973; Morley, in press). Snyder and Diesing (1977) have reported that
egotiators in crisis bargaining often did not set minimum goals at all.
Vhen minimum goals were set they were worked out late in the
egotiation.

Of course, sometimes what is acceptable can be determined only
vhen negotiation has shown what is available (Winham 1977). In
ther cases difficulties may arise because negotiators require direction
rom other members of a domestic group. For example, the American
Delegation in Paris found it difficult to get President Wilson's views
n questions such as: 'Are we agreed that the Alsace–Lorraine of 1871
hall be ceded? Or the Alsace–Lorraine of 1814? How will we meet a
lemand for the cession of the entire left bank of the Rhine?' In consequence negotiation proceeded from positions defined, generally, by a
Sritish or French draft. Wilson's ignorance of the European situation,
ombined with inadequate preparation, meant that influence
ittempts proceeded in the context of persistent obstruction, criticism
ind negation (Morley, in press).

But sooner or later negotiators have to establish a 'framework' within
vhich concessions can be made. For the moment it will suffice to note
hat some writers emphasize a process by which negotiators 'establish'
ind 'reconnoitre' a bargaining range (Douglas 1962) allowing
utcomes to be determined largely by perceptions of comparative
esolve. Others argue that the main problem is to get the negotiators to
ee issues in the same kind of way; to use the same 'strategic vocabulary'
Winham 1977; Zartman 1977). Zartman, for example, argued that:

'Rather than a matter for convergence through incremental concessions from specific initial positions, negotiation is a matter of finding
the proper *formula* and implementing *detail*. Above all, negotiators
seek a general definition of the items under discussion, concerned
and grouped in such a way as to be susceptible of joint agreement
under a common notion of justice.'   (Zartman 1977, p. 76.)

They may then begin the process of 'accommodation' and mov
towards a settlement.

Atkinson (1975) argues that one general objective of informatio
search should be to seek out 'links' between appropriate issues so tha
trade-offs can be made (and gives some useful examples of th
techniques which can be used). But issues can be linked only if the
have approximately equal value to the sides, and the more comple
the negotiation the harder it will be for negotiators to estimate wha
would be acceptable to the other side (Morley, in press). Th
negotiators studied by Balke, *et al.* (1975) were dealing only with fou
key issues (at the DOW Chemical Plant). They 'were confident tha
they understood their counterpart's policies, a belief based on years c
association and negotiation. Yet they were wrong.' (Balke, *et al.* 1975
p. 320.) Judgements of sample contracts showed that possibl
contracts were evaluated in different ways by different members of
team. Apparently, union negotiators had three management policie:
rather than one, with which to cope. Negotiators did not estimat
accurately the importance they attached to the issues, and change
policies 'from the evaluation of one contract to the next' (Balke, *et a.*
1973, p. 323).

Thus, inconsistency in cognitive judgements can prevent men c
goodwill from learning that, in principle, they see options in virtuall
the same way (Morley 1979).

Influence strategy and tactics

Put simply, negotiation is designed to get others to do what you wan
them to do − letting the other side have your own way. It is not th
same as persuasion, argument or debate:

> 'In 1971 a research unit prepared a long, reasoned and sophisti
> cated claim for the manual workers in their wage round in th
> automotive industry. It was not perfect, but it was a radica
> departure from historical precedents in manufacturing industry
> An eleven week strike followed. In 1972 a group of Scottis!
> engineering workers presented a claim for a substantial pa
> increase in these words, "Johnny wants a biscuit and he wants i
> now". A settlement acceptable to both sides followed immediately
> (Jenkins and Sherman 1977, p. 77.

Neither claim is typical, as Jenkins and Sherman admit. But the passage does demonstrate that whatever else goes on, bargaining is a matter of relative power. The decision making is decision making under risk, in which each party has something to lose (a stake) and something to gain (a prize). The stake and the prize are determined by the values and interests of the parties. It is these values and interests (or perceptions of them) that negotiators attempt to manipulate in the performance of their task. They also recognize that, to a greater or lesser extent, each is guided by his perceptions of the other's response: what is acceptable is a function of what is available.

Tactics may be classified as accommodative, coercive or persuasive, as offensive or defensive, attitudinal or situational. They may function to communicate firmness, or to influence estimates of the stake and the prize. They may also function to see that gains are made without resentment from the other side. The list of possibilities is almost endless.

Some treatments emphasize the kinds of arguments which are used in actual cases (e.g. Iklé 1964; Snyder and Diesing 1977). Others add behavioural advice such as: reply to a question with a question; phrase probing questions in general terms; paraphrase what has been said to you; and so on (Marsh 1974; Miron and Goldstein 1979). (I have not included a discussion on the use of threat. The omission is deliberate and intended to highlight the fact that experienced negotiators would prefer not to use threats unless absolutely necessary. For analysis of the reasons, and for more general discussion, see Morley (1979) and Morley and Stephenson (1977).) Snyder and Diesing (1977) deal with international negotiation; Walton and McKersie (1965) deal with industrial negotiation; and Miron and Goldstein (1979) deal with hostage negotiation. Here I want to outline some of the tactical moves suggested by Snyder and Diesing's rational bargaining module. In their own words:

> 'a rational bargainer builds redundancy into his messages. . . . He does not assume that the opponent "must know" what he is doing, but rather assumes the situation is pretty confused. Consequently he tries to send a message several different ways, always through a different channel, and keeps repeating the same theme. The purpose is to break through the resistance set up by the opponent's mistaken expectations and also to give him time to test, retest and adjust his expectations.'   (Snyder and Diesing 1977, p. 334.)

Snyder and Diesing define a strategy as a set of tactics. More precisely, a strategy is a set of tactics in a certain order. The key problem is how to select and sequence the elements in the mix.

Both Snyder and Diesing (1977) and Walton and McKersie (1965) argue that what is required is a more detailed analysis of the 'dilemmas' or 'choice points' created by conflicting goals such as 'winning' and 'avoiding war'. To illustrate the kind of thinking which is used I shall consider two very general problems which follow from a negotiator's concern to protect his 'reputation for resolve'.

(1) At any given point a negotiator may ask: Shall I make a concession? Shall I stand firm but signal flexibility? Shall I commit myself further to the position currently being advocated? The negotiator who decides to concede is in a position not unlike that of a girl on a date. If she 'bestows too early in the evening the only favours she is willing to give she is apt to find her partner acting on the expectation that more will follow if he persists' (Jervis 1970, p. 120). But if the negotiator stands firm he risks becoming committed to a position he cannot defend. According to Pruitt (1971) various cost considerations are made whether a negotiator makes concessions or not. And as time goes by pressures to concede and pressures to maintain demand increase. The result may be a progressive and deepening 'concession dilemma' in which negotiators fail to reach agreement even though their minimum goals are compatible. It is not clear how far this kind of argument can be taken (Winham 1977) but the dilemma is by no means confined to simple laboratory 'games' which demonstrate that concessions by one person raise the 'level of aspiration' of the other (Walton and McKersie 1965). In laboratory simulations it is clear that negotiators are sensitive to changes in the concession behaviour of their opponents. The research is complicated by the effects of time pressure, bargaining costs, and assumptions about the other's 'minimum terms': but it is quite common to find that (in some sense) the 'toughness' of A is negatively correlated with the 'toughness' of B. Further, extreme intransigence seems to decrease the likelihood of agreement in a variety of 'abstract' and 'realistic' games. For detailed analysis see Morley and Stephenson (1977) and Pruitt (1976).

(2) Can the negotiator stand firm but signal flexibility? Jervis (1970) argues that the answer is yes, but only when the signal is sufficiently ambiguous that his opponent cannot be sure what he is trying to say. The risk then is that the offer is not seen as an offer at all. An alternative approach is to place a number of specific illustrative proposals on the

table at the same time. Kissinger did this in the SALT talks at Helsinki, 1969 (Kalb and Kalb 1974, p. 116). According to Fisher (1969) this kind of move indicates that negotiators are 'searching for a solution and are not tied to one plan. And it will give an adversary several yesable propositions while encouraging them to work out more for themselves.' (Fisher 1969, p. 79.)

A second alternative may be to essay a GRIT strategy of the type recommended by Osgood (1960).* This policy has been widely misunderstood by social psychologists. However, the general ideas are fairly easy to state. A negotiator 'locked' in the concession dilemma (outlined by Pruitt) may be able to indicate flexibility *at minimum cost* by:

(a) Indicating his party would make specific moves designed to foster agreement;
(b) Announcing these moves in advance;
(c) Explaining that whilst the steps may seem *small individually* they would add up to something significant *if and when they were matched by concessions from the other side(s)*.

Negotiators may also have to take into account the skills their opponents bring to the task. Miron and Goldstein (1979) argued:

'One is constantly struck by the inadequacy of the problem-solving capability of the hostage taker. . . . Negotiation in hostage/barricade situations is simply the process of providing the criminal with a wider range of productive solutions to his problems.' (p. 18.)

What is recommended is in many respects the classic counsel for the avoidance of disaster: slow the momentum of events and maximize the time available to think calmly about the options (Snyder 1972); except that the negotiator must help provide a rationalization which helps the perpetrator to save face.

Decision making

The decision to negotiate   Selecting a strategy includes the decision whether to negotiate or not. It also includes the decision whether to push this issue. Batstone, *et al.* have reported that it is not unusual for 'leader' shop stewards to delay issues, waiting for the balance of power to shift in their favour. One steward commented: 'Management

---

* GRIT stands for graduated reciprocation in tension reduction.

aren't giving anything away at the moment . . . and we're in no position to fight. . . . We'll hold back, and then when management are crying out for production we'll hit them with this and they'll give it.' (Batstone, *et al*. 1977, p. 68.)

Identify constraints   Negotiation is a complex social activity in which actors wish to obtain favourable agreements, avoid disasters, ensure domestic support, educate constituents, return favours, maintain goodwill, improve bargaining reputations and so on and so forth (Walton and McKersie 1965; Morley and Stephenson 1977; Snyder and Diesing 1977). Walton and McKersie argue that goals may be grouped into four major categories: those concerned with distributive bargaining, integrative bargaining, attitudinal structuring and intra-organizational bargaining. In their view, there are different systems of activity concerned with each set of goals, each with 'its own internal logics and its own identifiable set of involvemental acts or tactics' (Walton and McKersie 1965, p. 4). For some discussion see Morley (1979).

Each objective may be regarded as a 'constraint' or an acceptable strategy, so that: 'the process of setting objectives is therefore a process of listing constraints and setting levels of aspiration or acceptability for each constraint' (Snyder and Diesing 1977, p. 363).

A further problem concerns the degree of generality at which goals are formulated (O'Leary 1973). However, an agreement will be reached only if it satisfies all of the constraints.

Decision-making crisis   Not many negotiations are conducted under a continual threat that one side will 'walk out' (Winham and Bovis 1978). Strategies are not revised very often (Snyder and Diesing 1978). It is often not clear until late in negotiation 'that agreement can only be had at a price hitherto thought unacceptable. It is thus necessary to think differently about the situation and to prepare for a difficult choice.' (Winham and Bovis 1978, p. 295.) Three aspects of this problem are worth further comment. They arise because negotiators, typically, negotiate in teams. They are also representatives of larger groups.

First, preparation and process are linked by virtue of the fact that they are the intra-group and inter-group phases of a complex decision-making task (Morley, in press). Team leaders have to bring together the right people at the right time (Winham 1977).

Second, negotiators must 'sell' a convincing package to domestic organizations outside the immediate negotiation; and third, the package requires marketing. The domestic organization will buy the package to the extent that it has been made attractive to them.

There are several determinants of the location of a negotiator's 'level of aspiration' or 'limit' at this stage. However, according to Magenau and Pruitt (1979) the two 'are frequently located at a particular alternative that has gained "prominence" in a bargainer's thinking' (p. 185). Put simply, negotiators need an obvious place to stop. Ideally, the 'end-point' should be intrinsically attractive (to the domestic organization), perceptually prominent (to emphasize this really is the end) and supported by one or more moral rules.

Psychologists have identified a number of different moral rules to which negotiators may appeal. Magenau and Pruitt list six principles related to 'social norms' of equity, equality, need, opportunities, equal concessions and historical precedent. The general style of their analysis is shown in the following passage:

'The equity principle is likely to be salient for people who are concerned about productivity. Equality is likely to be favoured when the negotiating parties are of equal status. The needs rule is most likely to be stressed when one or both parties feel responsible for the other's welfare. People who experience success, feel competent, or are otherwise in a good mood also place more emphasis on the other party's needs.'   (1979, p. 186.)

To anticipate a little it may be said that the needs rule comes to the fore when negotiators sustain a 'strong' relationship with their opponents. However, I shall be content for the moment to say that negotiators may appeal to several moral rules to support their case. The choice is affected by a large number of variables and is highly complex.

What promotes success in negotiation?

What promotes success in negotiation? That is, which kinds of *process* lead to a full and/or lasting agreement? (Winham and Bovis 1978.) Psychologists (and others) have attempted to answer this question in two ways. First they have looked at stages in successful and unsuccessful *negotiations*; at structure in the process. Second, they have compared *negotiators* with and without a track record of significant success. Each kind of research will be examined in turn.

Stages in negotiation

Negotiation which ends in agreement is not random; it is 'orderly and progressive in nature' (Douglas 1962). This much is common ground. There are, however, two very different accounts of the nature of the changes.

The first account emphasizes the similarity between negotiation groups and all other groups (Warr 1973): a group is a group is a group, and may be described in terms of broad categories of 'Interaction Process Analysis', developed by R. F. Bales. Using Bales's categories it has been demonstrated that the movement in negotiation groups is broadly the same as movement observed in laboratory problem-solving groups (Landsberger 1955). Furthermore, the greater the similarity the greater the extent to which specific items in dispute were resolved.

Contrastingly, the second account chooses to emphasize the *distinctiveness* of negotiation groups. Essentially, this asserts that negotiators face special problems precisely because they are acting as *representatives* of groups. They must, therefore, come to terms with two sets of forces: those produced by the demands of representing their parties and those produced by the demands of maintaining personal relationships with their opponents. Hence stages in negotiation may be represented as attempts to emphasize the interpersonal and interparty aspects of the negotiation task at different times (Douglas 1962; Morley and Stephenson 1977). The stages are notional (Pedler 1977) but are supported by empirical studies of negotiation in Great Britain and the USA.

Unlike the problem-solving process studies by Bales, negotiations begin (stage I) with a hard-hitting critique in which speeches are exceptionally long and participants emphasize the representative role they have to play. Further, 'to the extent that the contenders can intrench their seeming disparity in this period, the more they enhance their chances for a good and stable settlement in the end' (Douglas 1957, p. 73). What is achieved, apparently, is a definition of 'bargaining resolve' or 'strength of position' (Snyder and Diesing 1977; Morley and Stephenson 1977). That is to say, negotiators 'are assessing the form of the parties with respect to the particular course on which the present contest will take place' (Morley and Stephenson 1977, p. 288).

Subsequently (stage II), negotiators subordinate their representative roles whilst engaging in 'unofficial' behaviours, designed to 'reconnoitre the bargaining range' and give a more precise idea of the kinds of settlements which might be obtained. Finally (stage III), negotiators face a decision-making crisis similar to that described by Winham and Bovis (1978), and return to an emphasis upon their representative roles.

I wish to emphasize that stage II (reconnoitering the bargaining range) allows the bargaining relationship between the negotiators to come to the fore. Given a strong affective relationship between the negotiators, indiscretions and false moves will probably not be exploited by the opposition. For example, Batstone, *et al.* (1971) have pointed out that managers and stewards who had strong (rather than weak) bargaining relationships exchanged more information about the internal politics of their organization. Much of it prevented the other from 'getting into difficulty' or 'being conned' (p. 173). Further, there was a tendency to work out how goals could be achieved, and made to look legitimate in terms of previous agreements. That is, strong bargaining relationships helped participants each to attain their objectives.

Put another way, negotiation includes problem solving but problem solving which contains an irreducibly political component. The behaviour followed by the negotiators is behaviour which functions to sustain an exchange relationship between the participants. The stronger the bargaining relationship the greater the extent to which institutional goals will be modified by personal relationships between the negotiators themselves (Morley and Stephenson 1977).

It may, of course, be perfectly reasonable for negotiators to develop relationships of this sort. One shop-floor convenor interviewed by Batstone, *et al.* emphasized that it was important for him to have creditability with his firm: 'I don't want to be conned and neither does the company. It's wrong to screw (them); a section can go away, but the convenors have continually to go into management' (Batstone, *et al.* 1977, p. 172). But negotiators develop relationships of this sort at the risk of becoming isolated from the organization they represent.

Consequently, the stages identified by Douglas may be seen to have a compelling internal logic of their own. Usually, representatives may allow their unofficial problem-solving activity to come to the fore only when they have already been seen to do justice to the position of the party they represent.

In practice this means that negotiators must 'fail safe' with respect to the concession dilemma identified earlier. There may, of course, be more than one way to skin a cat, but we should note that in Snyder and Diesing's cases:

> 'the only good way to meet the loss-avoidance constraint on accommodation is *first* to establish a convincing image of firmness. If concessions are offered before this resolve image is established in the opponent's mind, he will not accept them . . . since he still believes himself the more resolved and therefore capable of winning the whole prize.'   (Snyder and Diesing 1977, p. 248.)

Further, conflict is the essential precursor of negotiation groups and, in a sense, negotiations begin with the expectation of a 'fight' (Morley 1979; Warr 1973).

The logic is compelling, but the risks are nevertheless severe. If stage I is to do more than demonstrate the 'gladiatorial skill of the contestants' it is essential that disagreement between the *parties* is not interpreted as *dislike* between persons (Douglas 1957, 1962; Landsberger 1955; Morley and Stephenson 1977). Apparently, experienced negotiators learn skills which mean that attempts by one party to influence and control the other do not undermine the personal relationships between the representatives.

There is much evidence to suggest that not to disagree means not to solve the problem. Research by Pruitt and Lewis (1975) is particularly relevant since it endorses some of the things I have been saying, whilst making it clear that several cautionary remarks may have to be made. Subjects played the roles of buyers and sellers attempting to reach agreement on the price of three minerals. The object of the exercise was to identify variables which increased bargaining efficiency, as defined by the *joint profit* which was obtained. Pruitt and Lewis (1975) concluded that:

> 'A period of conflict is often necessary before people look beyond the easy, obvious options in search of those that provide more joint profit. But we are endorsing conflict within the framework of an integrative mode of thought, the problem-solving orientation.'
>
> (p. 632.)

A 'problem-solving' rather than an 'individualistic' orientation was induced by instructions which contained four elements:

(1) View the bargaining exercise as a problem to be solved;
(2) play down the conflict nature of the task;

(3) make as much profit as you can; but

(4) be interested in the needs of the other company − essentially, the 'individualistic' orientation derived from (3) plus a statement that 'the needs and profits of the other negotiator are unimportant to you'.

Note that experienced negotiators rarely adopt an 'individualistic' orientation of this kind. Further, the 'problem-solving' orientation bears certain similarities to the 'high' bargaining relationship which has been described. By and large Pruitt and Lewis's (1975) data seem consistent with the position set out here. It is only element (2) which raises serious doubt about the mechanisms used. In practice the instruction may have had the effect of reducing the use of tactics such as 'positional commitments, threats, and arguments about why he should yield' (ibid., p. 627). Again, this is probably compatible with what has been said so far: but I do not want to gloss over possible sources of dispute.

Relevant empirical work is hard to find and it is quite clear that the analysis I have outlined is more appropriate to the early than the late stages of a developing programme of research.

In complex negotiations one practical way to reconnoitre the bargaining range is to establish trade-offs by linking issues (Winham and Bovis 1978; Atkinson 1975). Typically, subjects who reached agreements in the State Department Training Simulation described by Winham and Bovis traded concessions on one issue for concessions on what appeared to be totally unrelated issues. They did not attempt to negotiate issues one at a time, nor did they try to tackle all of the issues at once.

National teams (NTs) from successful 'runs' of the simulation gave more adequate information to their principals (inter-governmental teams, IGTs), particularly in stage III of the negotiations. Communications were not only more common but conveyed new information in the context of what had gone before. Participants were 'government officers trained to communicate in highly structured forms'. Nevertheless participants who failed to agree 'often conveyed information about issues of which the IGTs had little previous knowledge, with consequent confusion and loss of time' (Winham and Bovis 1978, p. 292). Apparently, taking the time to re-read some of the earlier NT−IGT communications was a simple step towards efficiency during negotiation.

One result was that teams reached agreement with each other because they found it easier to reach internal agreement amongst themselves, enabling the linking of 'roughly equivalent issues'. Once issues had been linked contingent agreements were reached and put to one side; the process was then continued until agreement was reached.

Successful teams were more efficient than unsuccessful teams in both a *positive* and a *negative* sense (Winham and Bovis 1977, pp. 291–2). Positive efficiency meant that they structured negotiation into a series of smooth steps and organized themselves to implement the 'plan'. Negative efficiency meant that they did not get bogged down with trivial issues. Nor did they debate at length positions which were 'manifestly unbalanced' or 'unequal' (see Winham and Bovis 1979 for a discussion of the concept of 'balance' in negotiation).

Comparisons of successful and unsuccessful negotiators

The suggestion that negotiation requires efficiency in both a positive and a negative sense receives support from two other sources. First, there are the studies by Rackham and Carlisle contrasting the behavioural tactics (Rackham and Carlisle 1978) and planning techniques (Rackham and Carlisle 1979)* used by 'effective' and 'average' performers. Second, there is the work of Karrass (1970) who studied the behaviour of 'skilled' and 'unskilled' performers in a laboratory negotiation of a legal damages case (Morley and Stephenson 1977, describe the negotiation as an example of a 'role-playing debate').

For Rackham and Carlisle being 'effective' means (1) being rated as effective by both sides: union and management, supplier and purchaser. In addition it means (2) having a good track record for reaching agreements and (3) agreements which stick. 'Average' negotiators are those who fail on (1), (2) or (3), or unknown quantities for whom no criterion data are available. For Karrass being 'skilled' means that the negotiator is rated as possessing desirable 'bargaining traits' by 'two of his managers'. Negotiators represented the buying and selling side of the American aerospace industry, and were defined as skilled if their 'trait score' lay above the sample median. Reassuringly, trait score was correlated with outcome. The more

* Rackham and Carlisle admit a number of methodological problems with their data but argue that nevertheless their work provides 'useful guidelines for the negotiator' (1979, p. 2).

skilled the negotiator, the greater the financial value of the settlement he obtained.

The positive side of efficiency   Positive efficiency refers to behaviour which creates 'a smooth flow of substance over time' (Winham and Bovis 1978, p. 291). The effective negotiators studied by Rackham and Carlisle (1978) were much more efficient, in this sense, than the average negotiators. They considered a wider range of 'options for action'. They were more likely to plan in terms of a range of possible outcomes than to 'plan their objectives around a fixed point'. They planned tactics using 'issue planning' rather than 'sequence planning'.

This last finding may seem rather confusing. Average negotiators placed very heavy reliance on verbalizations such as : First I'll bring up A, then B, then C, then D. This may sound as if it would help to keep negotiation running smoothly on the rails. However, as Rackham and Carlisle have pointed out:

'In order to succeed sequence planning always requires the consent and cooperation of the other negotiating party. The negotiator would begin at point A and then the other would only be interested in point D. This could put the negotiator in difficulty, requiring either a mental change of gear and approach to the negotiation in a sequence not planned for, or dogged insistence on the original sequence risking disinterest from the other party.'

(Rackham and Carlisle 1978, p. 4.)

Issue planning is planning which is independent of sequence.

The effective negotiator is also likely to use techniques suggested by Snyder and Diesing's rational bargaining module. For example, Rackham and Carlisle report a number of techniques used to slow negotiation down, reduce ambiguity and clarify communications. Effective negotiators tend to label behaviour, except disagreement e.g. 'Can I ask you a question?', 'If I could make a suggestion. . . .'). They hold back counterproposals rather than responding immediately. They talk about their own feelings rather in the style:

'I'm uncertain how to react to what you've just said. If the information you've given me is true then I would like to accept it, yet I feel some doubts inside me about its accuracy. So part of me feels happy and part feels rather suspicious. Can you help me resolve this?'   (Rackham and Carlisle 1978, p. 10.)

This may be one technique for ensuring that party disagreement does not turn into personal dislike. Another is to label threats as warnings (Snyder and Diesing 1977).

The effective negotiator tests his opponent's understanding and summarizes statements to a much greater extent than the average negotiator. This is consistent with the rational-bargaining module, and also reflects the effective negotiator's concern to make an agreement stick. Apparently 'average negotiators, in their anxiety to obtain an agreement . . . would prefer to leave ambiguous points to be cleared later, fearing that making things explicit might cause the other party to disagree' (Rackham and Carlisle 1978, p. 9). Consequently, their agreements ran into difficulty on implementation.

The negative side of efficiency   Most negotiators avoid antagonizing their opponents (Rackham and Carlisle 1978). The effective negotiator also avoids using 'irritators', phrases in which an actor says 'gratuitously favourable' things about himself.

Influence   Effective negotiators avoid putting counterproposals when the other party is preoccupied with stating his own case. They also use fewer reasons to back up each argument than unskilled negotiators. Apparently, negotiation is not usually a matter of advancing mutually supporting reasons to back up an argument. Rather, negotiation is an enterprise in which 'The poorest reason is a lowest common denominator: a weak argument generally dilutes a strong one' (Rackham and Carlisle 1978, p. 9). Consequently, the effective negotiator tended: 'to advance single reasons insistently, only moving to subsidiary reasons if the main reason was clearly losing ground' (Rackham and Carlisle 1978, p. 9). A further reason may perhaps be detected here to back up the 'internal logic' of the Douglas three-stage model of the process of negotiation.

Summary and discussion

'Negotiation is an enduring art form, its essence is artifice.' Artifice in the management of people through guile; artifice in the exploitation of potential force; artifice as problem-solving search; artifice in the construction of order and certainty in an uncertain and complex world. The art is working through the core processes of information interpretation, influence and decision making which are used. The product may be appreciated in various ways and at various levels. It is often difficult to say whether an artist is good or bad, and it is difficult to say why. Nevertheless, we can help people to produce paintings. They may even sell them to other people. In a crude sense we can see what the painting is worth. And we can see whether it keeps its value. Similarly, we can help people reach agreements. And we can see whether they keep their value.

People must, of course, be interested in producing pictures. There will be considerable conflict if they want to do something else instead, or if they want to paint a different kind of picture. This raises issues of some importance for negotiation. For example, collective bargaining is often seen as a rather conservative enterprise. To quote from Anthony's (1977) *The Conduct of Industrial Relations*:

> 'The accommodations arrived at in collective bargaining cannot penetrate too far into the attitudes and perceptions protected from contact in the inner recesses of the organization. The agenda is therefore concerned with differences over common ground, an area, which is judged to be suitable for collective determination.'
>
> (p. 230.)

Similarly, Chalmers and Cormick (1971) have argued that negotiating for agreements within the existing framework is simply a sophisticated way for Blacks to participate in their own oppression. They suggest that negotiating only the implementation of demands would change the balance of power, and go on to consider ways of making this possible.

Put simply, different people have different criteria for effectiveness. I have taken the view that negotiation includes the exploitation of potential force; that bargaining process is virtually inseparable from bargaining power. But there is a second kind of power which is much less evident in the psychological literature on negotiation: the power to define what constitutes an issue, and how that issue should be handled. The distinction can be overdone, but it is difficult to disagree with Edwards' contention that the positions stated represent two major and 'fundamentally opposing themes' (1971, p. 1).

Of course to study social skills, to talk about strategy and tactics at all we have to know the purpose of the social task. The work summarized in this chapter proceeds from the assumption that issues are, in some sense, agreed to be legitimate matters for joint regulation, and that the parties are *negotiating for agreement* (see p. 86). The position taken is similar to that outlined by Anthony, that is: 'Effectiveness may be judged to be increased by the removal of unnecessary and accidental obstacles in the way of the parties reaching a joint accommodation which they would both regard as a desirable end' (1977, p. 264).

Empirical research suggests that some of the ways in which effective negotiators set about structuring their behaviour to achieve ends of this kind may be identified. In particular, effective negotiators do not always adopt a 'legalistic' approach to negotiation in which it is no part of their job to do the other's work for him. Rather, the account given here emphasizes that:

(1) In many respects 'internal' bargaining within organizations and 'external' bargaining between organizations are to be explained in 'precisely the same terms' (Anthony 1977). In other words 'planning' and 'negotiation' are simply the intra-group and inter-group phases of a complex social decision-making task (Morley, in press).

(2) To a greater or lesser extent effective negotiators help others, colleagues *and* opponents, to come to grips with the 'core' problems of information processing, influence and decision making which define the job they have to do. Teaching the skills of negotiation is not just a matter of teaching a repertoire of tactics and moves apart from the particular cognitive and social problems which define the essentials of the negotiating process. Rather, the tactics must be seen to follow naturally from the bargaining process as it unfolds (Brandt 1972).

(3) Skill implies control and skill implies organization (Reason 1978). In this case the control is control over the values and interests which are at stake. The organization comes from understanding the nature of the negotiation task. *The skilful negotiator is one who through an understanding of the risks and opportunities associated with negotiation, and of the resources he can bring to bear, is able to take active and effective measures to protect or pursue the values and interests he has at stake.*

References

Abelson, R. P. (1973). The structure of belief systems. *In* Schank, R. and Colby, K. M. (eds). *Computer Models of Thought and Language*. San Francisco: Freeman.

Anthony, P. D. (1977). *The Conduct of Industrial Relations*. London: Institute of Personnel Management.

Atkinson, G. M. (1975). *The Effective Negotiator*. London: Quest Research Publications.

Balke, W. M., Hammond, K. R. and Meyer, G. D. (1973). An alternative approach to labor-management relations. *Admin. Sci. Q.* 18, 311–27.

Batstone, E., Boraston, I. and Frenkel, S. (1978). *Shop Stewards in Action: the Organization of Workplace Conflict and Accommodation*. Oxford: Basil Blackwell.

Boden, M., (1977). *Artificial Intelligence and Natural Man*. Hassocks: Harvester Press.

Brandt, F. S. (1972). *The Process of Negotiation: Strategy and Tactics in Industrial Relations*. London: Industrial and Commercial Techniques.

Chalmers, W. E. and Cormick, G. W. (eds) (1971). *Racial Conflict and Negotiations: Perspectives and First Case Studies*. Ann Arbor, Mich.: Institute of Labor and Industrial Relations, University of Michigan – Wayne State University and the National Center for Dispute Settlement of the American Arbitration Association.

Douglas, A. (1957). The peaceful settlement of industrial and intergroup disputes. *J. Conflict Resolution* 1, 69–81.

—— (1962). *Industrial Peacemaking*. New York: Columbia University Press.

Edwards, C. (1978). Measuring union power: a comparison of two methods applied to the study of local union power in the coal industry. *Br. J. Indust. Rel.* 16, 1–15.

Fisher, R. (1969). *Basic Negotiating Strategy: International Conflict for Beginners*. London: Allen Lane.

George, A. L. (1969). The operational code: a neglected approach to the study of political leaders and decision-making. *Int. Studies Q.* 13, 190–222.

Harrod, R. (1953). *The Life of John Maynard Keynes*. London: Macmillan.

Iklé, F. C. (1964). *How Nations Negotiate*. New York: Harper and Row.

Jenkins, C. and Sherman, B. (1977). *Collective Bargaining: What you always wanted to know about trade unions and never dared to ask*. London: Routledge and Kegan Paul.

Jervis, R. (1976). *Perception and Misperception in International Politics*. Princeton University Press.

Kalb, B. and Kalb, M. (1974). *Kissinger*. Waltham, Mass.: Little, Brown & Co.

Karrass, C. L. (1970). *The Negotiating Game*. New York and Cleveland: World Publishing.

Kinder, D. A. and Weiss, J. A. (1978). In lieu of rationality. Psychological perspectives on foreign policy decision-making. *J. Conflict Resolution* 22, 707–35.

Lall, A. S. (1966). *Modern International Negotiation: Principles and Practice*. New York: Columbia University Press.

Landsberger, H. A. (1955). Interaction process analysis of mediation of labor-management disputes. *J. Abnormal Soc. Psychol.* 57. 552–8.

Lockhart, C. (1979). *Bargaining in International Conflicts*. New York: Columbia University Press.

Magenau, J. M. and Pruitt, D. G. (1979). The social psychology of bargaining. *In* Stephenson, G. M. and Brotherton, C. J. (eds). *Industrial Relations: a Social Psychological Approach*. London: Wiley.

Margerison, C. and Leary, M. (1975). *Managing Industrial Conflicts: The Mediator's Role*. Bradford: MCB Books.

Marsh, P. D. V. (1974). *Contract Negotiation Handbook*. Epping, Essex: Tower Press.

Miron, M. S. and Goldstein, A. P. (1979). *Hostage*. Oxford: Pergamon Press.

Morley, I. E. (1979). Behavioural studies of industrial bargaining. *In* Stephenson, G. M. and Brotherton, C. J. (eds). *Industrial Relations: a Social Psychological Approach*. London: Wiley.

_____ (1981). Preparation for negotiation: conflict, commitment and choice. *In* Brandstätter, H., Davis, J. H. and Stocker-Kreichgauer, G. (eds). *Group Decision Making*. London: Academic Press.

Morley, I. E. and Stephenson, G. M. (1977). *The Social Psychology of Bargaining*. London: Allen and Unwin.

O'Leary, M. K. O. (1973). Policy formulation and planning. *In* Boardman, R. J. and Groom, A. J. R. (eds). *The Management of Britain's External Relations*. London: Macmillan.

Osgood, C. E. (1960). *Graduated Reciprocation in Tension Reduction: a Key to Initiative in Foreign Policy*. Urbana, Ill.: Institute of Communications Research, University of Illinois.

Pedler, M. (1977). Negotiation skills training – part 4. *J. Eur. Industr. Training* 2, 20–5.

Pruitt, D. G. (1971). Indirect communication and the search for agreement in negotiation. *J. Appl. Soc. Psychol.* 1, 205–39.

_____ (1976). Power and bargaining. *In* Seidenberg, B. and Shadowsky, A. (eds). *Social Psychology: an Introduction*. London: Collier Macmillan.

Pruitt, D. G. and Lewis, S. A. (1975). Development of integrative solutions in bilateral negotiation. *J. Personality Soc. Psychol.* 31, 621–33.

Rackham, N. and Carlisle, J. (1978). The effective negotiator – part I. *J. Eur. Industr. Training* 2, 6–10.

_____ and _____ (1979). The effective negotiator – part 2. *J. Eur. Industr. Training* 2, 2–5.

Reason, J. T. (1978). The passenger. *In* Singleton, W. I. (ed.). *The Study of Real Skills*. Lancaster: MTP Press, vol. 1.

Snyder, G. H. and Diesing, P. (1977). *Conflict Among Nations: Bargaining, Decision-Making and System Structure in International Crises*. Princeton University Press.

Steinbruner, J. D. (1974). *The Cybernetic Theory of Decision*. Princeton University Press.

Walton, R. E. and McKersie, R. B. (1965). *A Behavioral Theory of Labor*

Negotiations: an Analysis of a Social Interaction System. New York: McGraw-Hill.

Warr, P. B. (1973). *Psychology and Collective Bargaining*. London: Hutchinson.

Winham, G. R. (1977). Complexity in international negotiation. *In* Druckman, D. (ed.). *Negotiations: Social Psychological Perspectives*. Beverley Hills: Sage.

Winham, G. R. and Bovis, H. E. (1978). Agreement and breakdown in negotiation: report on a State Department Training Simulation. *J. Peace Res.* 15, 285–303.

Zartman, I. W. (1977). Negotiation as a joint decision-making process. *In* Zartman, I. W. (ed.). *The Negotiation Process: Theories and Applications*. Beverley Hills: Sage.

5 The supervision of working groups

NICHOLAS J. GEORGIADES and VANJA ORLANS

Introduction

Defining our terms

We have resisted the temptation of taking the potentially narrow definition of 'supervision' which might be implied in our title. For some, 'supervisor' may be synonymous with 'foreman'. Alternatively, the supervisory role may apply to any leadership role within the structure of the organization, where a formally designated leader is in charge of a working group.

There are a number of problems in adopting the narrower definition: one set of problems relates to the degree of understanding which may be gained by viewing the role of foreman in isolation; more emphasis tends to be placed on the differences between the foreman and the rest of management, rather than on the similarities which may exist. Another set of problems arises when considering the issue of training in social skills.

Our broader orientation allows for a consideration of the similarities which exist across management functions; it allows also for a consideration of research which has not been limited to the foreman role. Most importantly, we believe that the social skills necessary for the management of a working group may be applied to all management functions.

The relevance of social skills

The skills used in social interaction both for individuals and for groups continue to gain greater emphasis in industrial and professiona raining courses. This may be explained in one of two ways. First, the rowth of industrialization, the rapid changes in technology, the ncreasing size of organizations and the increase of multiplicity of unctions within organizations have created an extremely complex nvironment with which the individual has to cope. Training in social kills may help the individual to deal more effectively with the esulting feelings of ambiguity, conflict, isolation and constant hange. Secondly, our understanding of what managers and superisors actually do has been greatly enhanced during the past 20 years. IcCall, *et al.* (1978) examine twenty-three studies of managerial ctivity from 1951 to 1976. The studies, conducted using different nethods, different sample size and managers at different levels in neir organizations show a remarkable level of replication. The most nportant results for this chapter are those that suggested that:

1) A manager's work is fragmented, brief and subject to interruptions and discontinuity. For instance, a study of foremen showed one incident every 48 seconds, with individual foremen averaging from 26 seconds to 2 minutes per incident. (Guest 1956.) Similarly, Mintzberg (1970) reported that half of a Chief Executive's activities lasted less than 9 minutes each and only 10 per cent of all activities last an hour or more.

2) A manager's work is primarily oral. For instance, studies of foremen have shown that between 28 and 80 per cent of their time is spent in verbal communication. (See for example Guest 1956; Hinrichs 1964; Jasinski 1956; Ponder 1957.) Managers at higher levels spend up to 90 per cent of their time in verbal communication, with 65–75 per cent being most typical (see *inter alia* Burns 1957; Dale and Urwick 1960; Horne and Lupton 1965). One chief executive, for example, spent 42 hours of the working week making 'verbal contact' (Mintzberg 1970). It is not recorded what proportions were spent in talking and listening! Most verbal interactions are face to face through formal or informal meetings, and as rank increases more time is spent in scheduled meetings with groups of people. Stewart (1967) found 32 per cent of the managers' time was spent in meetings with one other person and 34 per cent was spent in meetings with two or more.

3) Managers make use of a large number of contacts. Studies across a range of management levels have shown that 26–28 per cent of a manager's time was spent with subordinates (Kelley 1964;

Stewart 1967) and that 32—46 per cent of their *contact time* wa
spent with subordinates (Kelley 1964; Ponder 1959). Interaction
with peers are also important. Through low to middle levels o
management, for instance, contacts with peers generally consis
of about a third to a half (or more) of a manager's interna
(within the organization) contacts (Blau 1954; Burns 1954
Lawler, *et al.* 1968).

These three sets of data, highlighted by McCall, *et al.* (1978)
indicate the extent to which the reality of the supervisory role impose
the need for superior social skills. Conversely, it has been argued b
Guest (1956) that the actual demands of the job make the implement
ation of many 'human relations' strategies virtually impossible an
that therefore 'you could dispose of all leadership training course
without anyone knowing the difference'.

We do not hold to this view. We would suggest that these dat
emphasize the extent to which human interaction is a central proces
in a manager's work and how, therefore, the skills required must b
'over-trained' or learned to high levels of performance so that thei
implementation becomes 'second nature' to the performer.

Interpersonal skills

There is a paucity of taxonomic approaches in the manageria
training literature to the development of interpersonal skill. On
notable exception would perhaps be the work of Rackham *et a*
(1971). However, we have been more impressed by the work derive
from the field of counselling as examples of a more systemati
approach to the development of interpersonal skill; in particular wit
the work of Egan, derived as it is from the work of Carkhuff (1972
and his co-workers on the systematic development of skills training fo
the 'helping professions'. Egan has been responsible for developin
systematic programmes of skill development of value to managers. O
particular interest are his presentations in 1976 and 1977. Ega
(1977) suggests that his approach is an invitation to the participant t
ask the questions: 'Just how well do I communicate with the people i
my life? Do I want to improve my ways of relating to others? If so, ho
can I go about it?' (Egan 1977, p. 3.)

While not directly identifying a taxonomy of skills to be develope
by the trainee, such a taxonomy is clearly implied in his writing

He identifies his learning approach as *systematic* and by that he means 'learning one skill at a time by means of a step-by-step process' (Egan 1977, p. 5). The sequence of skills presented includes each of the following:

(1) Development of self-awareness, to facilitate greater understanding of one's effect on others and to facilitate self-disclosure.
(2) Development of a self-disclosing capacity in one's relationships with others. Of particular relevance here is the work of Jourard (1964) and Jourard and Lasakow (1958). Culbert's (1968) paper is an interesting review article of these and other authors' works in this area.
(3) The ability to avoid being vague in communication. The development of the skill of 'concreteness' in interpersonal interaction.
(4) The skills of listening and responding, which would include paying attention to others; 'hearing' what others say with their bodies, their gestures and their faces; understanding accurately what others are thinking and feeling; communicating to others that you do understand or that you are trying to understand them.
(5) The skills of challenging and confronting the significant other. Egan (1977) defines confrontation as 'anything that invites a person to examine his or her interpersonal style — emotions, experiences and behaviours and its consequences (for instance, how it affects others) more carefully'. Egan (1977) cites Berenson and Mitchell's 1974 classification of five different kinds of interpersonal confrontation: namely, information confrontation, experiential confrontation, strength confrontation, weakness confrontation and encouragement to action. Each of these skills, suggests Egan, may be developed by extensive skill training.

Implicit within each of these broad categories of social skill is an increasingly complex system of verbal interaction designed to aid the relationship between two parties. Timmons (1974) in his paper on the formation and development of the entrepreneurial team emphasizes how vital he believes skills of these kinds to be. Without apology he talks of the need of successful entrepreneurs to possess 'helping skills'. He says: 'It is difficult to think of a successful team in organizations or athletics, for example, whose members don't feel they are helped a great deal by their team-mates. . . . Giving and receiving help are skills which are increasingly being acknowledged as important aspects of effective managing.' Kolb, *et al.* (1971) support that contention. The skills articulated by Egan, derived essentially from the literature of the helping professions, are central to these social skill training developments.

Factors affecting the management of groups

We identify two major factors which affect the management of a working group: group processes and group tasks, and the leadership role. We shall distinguish between the social skills used at these two levels.

Group processes and group tasks

In general, groups may be either formal or informal; a formal or purposive group is one that is created deliberately to accomplish something. Formal work units whose boundaries or job responsibilities appear on the organization chart belong to this category. Within this work group, or overlapping with it, spontaneous or informal groups may emerge. These informal groups may work either for or against organizational goals.

Both formal and informal groups develop norms or sets of rules to which the group members are expected to conform. An individual belongs to a number of different groups at any one time, and conflicting norms may arise. There is usually considerable pressure on the individual to conform to group norms.

Much research interest has been shown in what has been termed 'group cohesiveness'. Cohesiveness has been variously defined as morale, task involvement, feelings of belongingness, shared understanding of roles and good teamwork. In general, as the cohesiveness of a work group increases the overall level of member conformity to the norms of the group would also be expected to increase (Festinger, *et al.* 1950). However, high levels of cohesiveness may or may not lead to higher productivity. If cohesiveness in the group is based on interpersonal rewards, productivity will probably be lower than if it had been based on a shared commitment to the task (e.g. Back 1951). In relation to social skills training this is an important distinction.

In attempting to generate a shared commitment to the task, group structure becomes important; the demands for refined and complicated group structures vary from one task to another and may have important implications for group member satisfaction and cohesion. Rosen (1973) emphasizes the importance of task—technological variables which create significant differences between formal groups. The variables define the very nature of the tasks done by each work unit. Task—technological variables include:

(1) Pressure for accuracy — this may come both from the employer and outside organizations.
(2) Pressure for speed — e.g. the relative importance of deadlines.
(3) Technological autonomy and task requirements — that is, how and what to do, and when to do things.
(4) Interaction opportunities — between individuals as well as across departments and disciplines.
(5) Role interdependence — some tasks can be performed only if work-unit members work together as a team, producing high role interdependence.

Very often these task—technological variables are outside the control of the group or the immediate supervisor.

While situational constraints must clearly be taken into account, some research has attempted to understand the nature of the relationship between group interaction processes and effectiveness. Hackman and Morris (1975) suggest that one key to understanding the effectiveness of small groups is to be found in the interaction process that takes place among members as they work on a task; at one extreme, for example, group members may work together so badly that members do not share uniquely-held information that is critical to the problems at hand; hence the quality of the group outcome will surely suffer. On the other hand, group members may operate in great harmony, with the comments of one member prompting quick and sometimes innovative responses in another, which then leads a third to see a synthesis between the ideas of the first two, and so on. In this case a genuinely creative outcome may result.

This raises the issue of group communication patterns. In a sense, all communication within groups is between individuals and is, therefore, interpersonal communication; this relates to our earlier emphasis on the importance of interpersonal skills.

There are a number of communication patterns which may be observed in a group. One is the relative frequency and length of communication acts: who communicates, how often, and for how long.

A second pattern is who communicates with whom; this is often helpful in pinpointing conflicts, or for increasing the group members' understanding of how they are relating to one another. A third pattern relates to who 'triggers' whom and in what ways; for instance, when one group member speaks does another always speak next, even if the remarks are not initially directed at him? Triggering may reflect either support ('attaboy') or a desire to undo the point ('yeabut') Schein (1969) quotes a business man as saying that in group discussions in his company it takes at least three 'attaboys' to undo the damage of one 'yeabut'. Another type of triggering is when one group member interrupts other members. Generally, high-authority members feel freer to interrupt low-authority members than vice versa.

In structuring its communication patterns a group develops a communication network which determines the amount and type of information a group member will receive from other members. In a one-way communication network the listener is entirely passive; communication effectiveness is determined by how the messages are created and presented. A one-way-with-feedback communication procedure has been termed directive or coercive communication (McGregor 1967); it is coercive because no provision exists for mutual influence or exchange. Two-way communication is a reciprocal process in which each member initiates messages and tries to understand the other member's messages. Social skills in relation to communication will be considered in a later section.

We turn now to a consideration of these processes in relation to task effectiveness. Hackman and Morris (1975) suggest that the effect of group interaction on group effectiveness is not direct, but instead operates by affecting three 'summary variables' that determine how well a group does in its task. These variables are:

(1) The level of effort the group applies to carrying out its task.
(2) The adequacy of the task performance strategies used by the group in performing the task.
(3) The level and appropriateness of the knowledge and skills brought to bear on the task by group members.

Hackman and Morris suggest three approaches which are designed to create conditions that encourage group members to engage in serious explorations of their norms about strategy when there is reason to believe that existing norms are not optimal for the task at hand.

The first two approaches: diagnosis–feedback (Jackson 1965) and process consultation (Schein 1969) require group members to address directly group norms about performance strategy and generally require the presence of an outside consultant. Interpersonal and group relations skills on the part of both the leader and the subordinates are particularly important in relation to these two approaches.

The third approach – task-design – deals less directly with norms, and minimizes the role of outsiders; the objective is to get potentially task-effective patterns of behaviour underway. However, these behaviours may not necessarily be internalized into the existing group norms, producing little consequent increase in performance level. The first two approaches have the advantage of placing responsibility for change at the feet of group members, thus generating greater commitment. Further, group-relations skills which are acquired may well be used for a variety of tasks.

In practice a combination of the three approaches might well be most effective. Research by Hackman and Oldham (1976) suggests that an individual's task motivation is often enhanced when jobs include much of the following:

(1) Skill variety: the degree to which the individual does a number of different things on the job using his or her valued skills.
(2) Task identity: the degree to which he or she does a whole and visible piece of work.
(3) Task significance: how much the results of work on the job will affect the psychological or physical well-being of other people.
(4) Autonomy: the personal initiative and discretion the individual has on the job.
(5) Feedback: the degree to which the person learns while working how well he or she is doing.

Task–technological variables of the type outlined earlier might seriously limit the degree to which these dimensions may be incorporated into a job; a notable exception is that of feedback.

Group relations skills

Our emphasis here is on those skills which may be used both by managers and subordinates in a group. The effectiveness of these skills depends largely on the interpersonal competence of individual group members.

A later section deals specifically with social skills of the leadership role. Group skills include:

(1) The ability to understand the nature of group processes, their complexity and their interrelatedness.

(2) Understanding the relative importance of the task and the effects of situational constraints.

(3) Diagnosing which group processes lead to greater task effectiveness.

(4) Contributing towards creating an atmosphere of openness and trust.

(5) The ability to diagnose group communication processes.

(6) Facilitating better communication within the group by attempting to establish a co-operative group climate that encourages participation of all members, and by promoting group norms that foster the feeling that a member's ideas and views, no matter what his authority level, are of real interest to other group members.

(7) Being able to manage conflict constructively within the group. This includes working towards a common set of norms and values in relation to how conflict should be handled. It also includes a consideration of strategies which may be used to resolve conflict, e.g. win—lose, problem solving and negotiation.

(8) Identifying and clarifying group goals, stating them in operational terms, and assessing their appropriateness and effectiveness.

(9) Being able to recognize and use all available group resources.

Leadership

Most approaches to the study of leadership have tended to emphasize prescription rather than description. In addition, the focus has tended to be somewhat narrow, concentrating on the leader and his subordinates. The process of leading takes place in a far wider environment including superiors, contacts outside of the organization and environmental variables. We argue that it is necessary to concentrate on what leaders, managers or supervisors *do* before we can identify the range of social skills used and can therefore help them to perform more effectively.

Below we review the major approaches to the study of leadership and attempt to identify some of the more important social skills in the leadership role.

Whereas early work on leadership attempted to isolate personality characteristics which distinguished successful from unsuccessful leaders, the initial hope that leaders shared common characteristics across situations was not borne out. Several studies emphasized instead the 'style' a leader uses in dealing with subordinates. Many different labels have been generated to describe essentially two styles: task-oriented, and person-oriented. Since the early 1950s numerous studies have been conducted attempting to determine the relationship between these two dimensions and leadership effectiveness. Preeminent amongst these were the studies conducted by research workers at Ohio State University and at the University of Michigan.

The Ohio State studies used the leadership dimensions labelled 'initiating structure' and 'consideration' and conducted their investigations using questionnaires completed by the leader, the leader's subordinates, peers and the immediate supervisor. They initially sought to determine the most effective leadership style. The findings indicated that a mix of initiating structure and consideration by the leader achieved the highest levels of effectiveness. The results showed that no single style emerged as being 'best'. For example, in some studies high initiating structure and high consideration were associated with high performance and worker satisfaction. Other studies showed that this style profile produced some dysfunctional effects. In particular, in a study of manufacturing firms consideration was found to be positively related to absences and negatively related to performance ratings of the leader by his superiors, and initiating structure was associated with lower worker satisfaction.

The University of Michigan studies used the dimensions labelled 'job centred' and 'employee centred' and conducted their investigations using questionnaires completed only by subordinates. Employee-centred and job-centred styles resulted in productivity increases. However, job-centred behaviour created tension and pressure that resulted in lower satisfaction and increased turnover and absenteeism. The main conclusions from these studies would seem to be that leadership style effectiveness should not be evaluated solely by productivity measures but should include other employee-related measures, such as satisfaction. In this framework, the supporters of the Michigan approach might conclude that employee-centred leaders' behaviour would be most appropriate and effective.

Both groups of studies suggest that the two dimensions, task oriented/people oriented, represent reliable phenomena in the measurement of leadership behaviour. However, at least three concerns emerge for the student of the literature in general and for the trainer in particular.

First, are the two dimensions truly independent? Various studies have shown them to be positively correlated, uncorrelated and negatively correlated. One explanation of this apparent confusion may be related to the particular form of questionnaire used by the researchers. See, for instance, the work of Weissenberg and Kavanagh (1972), who suggest that subordinates are unable to perceive the two dimensions as separate.

Second and related is the concern expressed by McCall (1977), who suggests that while these studies were initially intended to reflect the behaviour of leaders the styles are most commonly measured by paper-and-pencil questionnaires reflecting self- or others' reported perceptions rather than the actual behaviour of leaders.

Finally, since not only are the results confusing but also because the data-collection methods are of dubious behavioural precision, the trainer, intent upon designing social skill programmes to deal with these specific dimensions, is severely hampered. While the research programmes have produced many papers of outstanding scholarship and have undoubtedly helped secure the tenure of many excellent academics, they have helped little to provide the precision needed to build social skill training programmes, save at the most global level of understanding.

Argyris (1976) has attempted to move away from an analysis of 'good' or 'bad' leadership behaviour, and emphasizes instead the learning or re-education of managers; his view is that 'effective leadership and effective learning are intimately related'. Argyris distinguishes between 'single-loop' and 'double-loop' learning; double-loop learning uses not only an understanding of how well goals are being achieved but also confronts the validity of the goal or the values implicit in the situation. Essentially, Argyris argues for a shift in values; he proposes a model of double-loop learning based on the values of valid information, free and informed choice and internal commitment to the choice together with constant monitoring of the implementation. Though compelling, Argyris's approach has not been validated extensively.

A somewhat narrower perspective has been adopted by Vroom

(1976) and Fiedler (1967, 1976). Vroom focuses on the development of a normative or prescriptive decision-making model for leaders (Vroom and Yetton 1973). He believes that the behaviour of the leader is determined by attributes of the leader himself and attributes of the situation he encounters. An effective leader is neither universally autocratic nor universally participative but may utilize either approach in response to the demands of a situation as he perceives them. In a training context individuals are encouraged to compare their decision-making styles with the normative solutions based on a careful situational analysis. Vroom and Yetton identify five behavioural styles ranging from autocratic to participative, to group-decision or consensual. In addition, they identify three general criteria of effective decision making:

(1) The objective quality of the decision.
(2) The time required to reach it.
(3) The degree to which the decision is acceptable to subordinates.

In addition, there are various attributes of the decision-making situation, e.g. do subordinates have the information required to generate a high-quality decision? Or are subordinates likely to be in disagreement about preferred solutions?

Two problems may be identified with this approach. One problem relates to the narrow perspective; for instance, what about the degree to which the decision is acceptable to superiors? Second, it assumes that a leader is in a position voluntarily to change his behaviour.

Fiedler emphasizes the effective diagnosis of the situation in which the leader will operate. He suggests that a leader can change his leadership behaviour only in situations in which he has a great deal of control; the emphasis is consequently on changing the situation. Fiedler has identified one measurable leader characteristic which is claimed to relate consistently to effective leadership. This trait, esteem for one's 'least preferred co-workers' (LPC) is predicted to interact (either positively or negatively) with three situational characteristics: the structure of the task, the quality of the leader–members relationship and the formal 'position power' of the leader. Fiedler's research provides support for this complex theory; other research, however (e.g. Graen, *et al.* 1970), does not. As yet, the theory seems inadequate for the selection or training of leaders. The major strength of the theory is its emphasis on the interactive nature of the leadership process; however, no adequate taxonomy of situations yet exists.

One major problem with much of the leadership research relates to the definition of leadership; a leader may be the formal role occupant but leadership may be distributed throughout a group, where it may be defined as a 'process of influence'.

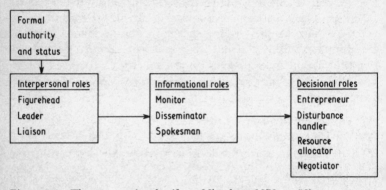

Figure 5.1   The manager's roles (from Mintzberg 1973, p. 59).

Most approaches emphasize what leaders *should* be doing, rather than what they *are* doing. Leaders, managers and supervisors are subjected to a variety of pressures: there is often not time to consider all the available data before making a decision. Very broadly, the leadership role encompasses the task functions and the maintenance functions. The task functions relate to task goals, whereas the maintenance functions relate to the working relationship within the group. Mintzberg (1973) has been specifically concerned with what managers do; he identified ten different managerial roles which cover both task and maintenance functions. He emphasized the interrelated nature of these roles as well as their contingency (Fig. 5.1). The interpersonal roles arise directly from the manager's formal authority. By virtue of his interpersonal contacts, the manager emerges as the 'nerve centre' of his organizational unit. He may not know everything, but he typically knows more than any member of his staff. Information in turn is the basic input to decision making.

Mintzberg emphasizes the somewhat chaotic nature of the manager's job. However, Drucker (1954) saw the manager as 'composer and conductor' in control of his orchestra. The reality as perceived by Sayles (1964) is:

'(The manager) is like a symphony-orchestra conductor endeavouring to maintain a melodious performance in which the contributions of the various instruments are co-ordinated and sequenced, patterned and paced, while the orchestra members are having various personal difficulties, stage hands are moving music stands, alternating excessive heat and cold are creating audience and instrument problems, and the sponsor of the concert is insisting on irrational changes in the programme.'

Social skills and leadership

Although much of the leadership research is as yet inconclusive, it does suggest a number of skills which may be useful in the context of the leadership role. Mintzberg's research in particular lends support to our earlier emphasis on interpersonal skills. The group-relations skills which we have outlined are important in relation to the group of subordinates as well as to groups of superiors, peer groups and groups outside the organization. A number of specific leadership skills would include:

(1) The ability to distinguish between 'leader' and 'leadership'. Leadership implies that one person is influencing other group members. A designated leader implies that one person is in charge of the group and has been given authority to exert influence within it. In any effective group the designated leader is not the only member to engage in leadership behaviour; any other member can be the leader when he or she influences the others to help the group reach its goals.

(2) Being able to distinguish between task and maintenance functions, recognizing when a task orientation or a maintenance orientation will be more productive. Task functions include:
   (a) information and opinion giving;
   (b) information and opinion seeking;
   (c) giving direction;
   (d) summarizing;
   (e) co-ordinating;
   (f) diagnosing;
   (g) reality testing;
   (h) evaluating.

Maintenance functions include:
 (a) encouraging participation;
 (b) harmonizing and compromising;
 (c) relieving tension;
 (d) facilitating communication;
 (e) evaluating the emotional climate;
 (f) observing process;
 (g) listening actively;
 (h) building trust;
 (i) solving interpersonal problems;
 (j) setting standards.

(3) Understanding different decision-making methods, e.g. decision by consensus, majority vote, authority.

(4) Analysing situational and human factors in decision making.

(5) Defining a problem effectively, using problem-solving procedures, deciding on and implementing a strategy to solve a problem, and evaluating the success of this strategy.

(6) The ability to build and develop an effective team. The ideal effective team as outlined by Johnson and Johnson (1975) is one which:

> 'has clear, cooperative goals to which every member is committed; accurate and effective communication of ideas and feelings; distributed participation and leadership; appropriate and effective decision-making procedures; productive controversy; high levels of trust, acceptance and support among its members, and a high level of cohesion; constructive management of power and conflict, and adequate problem-solving procedures'.

We stress that this is very much an 'ideal' and that a large number of variables may militate against its achievement.

Training

In this section we review supervisory and managerial training to date, before outlining some possible developments.

Supervisory training

Towards the close of the Second World War the Training Within Industry (TWI) programme was imported from the USA to the UK.

The programme included human relations training, elementary work study, methods of teaching and instruction and the problem of safety. Although TWI had been criticized as consisting of ineffectual stimula with little long-term effect on supervisory behaviour, it nevertheless made the experience of adult training familiar to many managers and supervisors for the first time. The TWI programme made two assumptions: the assumption of common skills, and the assumption that training could be packaged. Subsequent developments saw the proposal of a four-stage approach to supervisory training outlined by the Central Training Council in 1966. These four stages consisted of:

(1) Job analysis.
(2) Appraisal of individual knowledge and skill.
(3) Planned training for individuals or small groups.
(4) Follow-up of job performance after training.

A similar approach was adopted by the Engineering Industry Training Board (EITB) and the Iron and Steel Industry Training Board. This represented an attempt to move away from the common-skills approach of the TWI programme; however, the assumption that a programme could be packaged still remained. The EITB approach in particular is based on the assumption that individual managers know what their supervisors should be doing and that the chief problems are those of classifying and standardizing their beliefs. As Thurley and Hamblin (1967) point out: 'it may be dangerous for managers to assume that their own expectations of supervisory behaviour necessarily correspond with either the needs or the demands of the immediate or long-term production system'.

As a result of the demise of the British Institute of Management (BIM) Certificate in Foremanship and Supervision, a number of interested parties were invited to form the National Examinations Board of Supervisory Studies (NEBSS) in 1964. The board has developed a number of courses for the training of industrial supervisors. Underlying the board's main objectives was the assumption that training needs could be identified by higher-level managers and that training could be packaged. The Industrial Training Act distinguishes between training and education; the NEBSS courses appear to relate to education rather than training, which raises the question of which should come first.

Although current trends indicate a growing awareness of the

problems inherent in the available training programmes, the solutions to these problems are being viewed in terms of alterations to the existing programmes. Many basic assumptions on the nature of the supervisor's job and the training needs of supervisors are not being questioned. Also, the values on which the training courses are based are not made explicit. These values by and large reflect the values of more senior management, and the training courses might be viewed as an attempt to shift the values of supervisors (who very often are promoted from the shop floor) to bring them more into line with management thinking.

Management training

Within the higher levels of management the emphasis has traditionally been on education rather than training. A large number of courses are offered by Polytechnics and Universities leading to the Diploma in Management Studies, the MBA, etc. In relation to training, an approach which is still popular consists of the traditional management trainee scheme where young managers spend varying lengths of time in the different departments of an organization, presumably 'learning the ropes'; this may be regarded as induction rather than training.

Current trends emphasize management training *and* development, reflecting a need to view training as a continuing process. Training courses for managers tend largely to be based on developments in the USA; the courses typically include decision-making strategies, creative problem-solving, team building and group relations training. Although managers are encouraged to focus on personal and organizational goals, little data is yet available as to whether these goals are achieved, or whether training makes any difference.

One problem surrounding management training relates to the fact that courses are generally emphasizing what managers should be doing without taking into account what they are doing. Unless we know what managers actually do, it is difficult to design courses to help them perform more effectively.

There have been several approaches to the analysis of the management function. However, as we pointed out in our earlier section on leadership, only fairly recently have a number of researchers begun to emphasize the actual nature of the manager's job; as a result a number of important skills have been identified, all of which must be practised in a somewhat chaotic environment. These skills include developing peer relationships, carrying out negotiations, motivating subordinates, resolving conflicts, establishing information networks and subsequently disseminating information, making decisions under conditions of extreme ambiguity, and allocating resources (Mintzberg 1973). There are as yet few training courses which are based on these developments.

As with other types of training, there is always the danger that an organization will place greater emphasis on being seen to be training and keeping up with current developments than on examining the values and assumptions on which training is based, and developing a training policy to which members of the organization are committed. For example, successful implementation of Management by Objectives includes an understanding of, and a commitment to the values of participation, openness and trust.

It is not uncommon to hear of organizations which send vast numbers of managers on training courses in Management by Objectives to fulfil the injunction from on high that 'as from the 1st April all managers will implement the XYZ M. by O. scheme'.

Sensitivity training and role play: their part in social skills training

No two training techniques have aroused more controversy, nor stimulated more intense scholarship, than sensitivity training (the T-group) and role-playing techniques. No summary article of this kind would be complete without reference to both. However, space precludes the detailed analysis which both of these training techniques deserve.

Sensitivity training: The T-group In 1946 the National Training Laboratories, based in Bethel, Maine USA, asked Kurt Lewin to help them develop and present a training programme for community leaders. The leaders were brought together to discuss various social problems. Observers of these sessions fed back what they had observed to the participants. Thus the true subject matter of the training session became the behaviour of the trainees. Campbell, *et al.* (1970) note that since those original experimental learning events there have developed many variations of the original NTL groups. Despite these wide variations, these authors felt that certain objectives were common to various sensitivity training techniques:

(1) To give the trainee an understanding of how and why he acts towards other people as he does and of the way in which he affects them.

(2) To provide some insight into why other people act the way they do.

(3) To teach the participants how to 'listen', that is, actually hear what the other people are saying rather than concentrating on a reply.

(4) To provide insights on how groups operate and what sorts of processes groups go through under certain conditions.

(5) To foster an increased tolerance and understanding of the behaviour of others.

(6) To provide a setting in which an individual can try out new ways of interacting with people and receive feedback as to how these new ways affect them (Campbell, *et al.* 1970, p. 239).

There have been innumerable studies of the effectiveness of group training methods. Gibb (1974) cites over 300 references on the effects of group training. The arguments surrounding the effectiveness or otherwise of the methodology are many and complex, and not entirely subject to normal standards associated with objective evaluation.

Thus two recent industrial psychology textbooks (both of American origin) reviewing the identical summary article of T-group research conclude with diametrically opposing views (Landy and Trumbo 1980; Ivancevich, *et al.* 1977). Perhaps the most thorough review of the entire methodology and its evaluation, by Blumberg and Golembiewski (1976), concludes as follows:

(1) By far the majority of those taking part in T-groups report that it was a meaningful experience or at least a pleasant one.

(2) Only a fraction of 1 per cent report persisting or substantial negative experiences.

(3) On balance the evidence favours the view that learning or change does occur as a result of the experience and that the learning is of the kind intended. For some training objectives, particularly those related to small-group dynamics, the T-group methods seem particularly potent.

(4) Listening skills, emphasized so strongly in the Egan approaches

to social skills training, also seem to be considerably enhanced by participation in T-groups. Blumberg and Golembiewski (1976) say: 'An impressive array of studies support the positive effects of T-group experiences on listening.'

(5) In the area of the improvement of interpersonal perception — the ability to see ourselves as others see us, and to see others as they see themselves — the research evidence is inconclusive or negative. As Campbell and Dunnette (1968) in an extensive review article conclude: 'the studies incorporating a measure of how well an individual can predict the attitudes and values of others before and after T-group training have yielded largely negative results.' There is slight evidence for an improvement in self-perception, though at least one major study (Gassner, *et al.* 1965) provides mixed findings about the effects of training on self-perception.

However, despite a vast number of research studies the methodological problems associated with research on the T-group and related techniques are manifold. For a summary of these difficulties see Harrison (1971). There can be little doubt, however, that for many individual participants the T-group experience is associated with a 'warm glow' of personal confirmation.

While some studies, particularly those of Lieberman, *et al.* (1973), have indicated a 'casualty rate' of 7.5 per cent (where a casualty was defined as someone who had suffered some sort of psychological injury that lasted weeks or months) a more recent replication by Kaplan, *et al.* (1980) indicates that the 'casualty rate' was closer to 2 per cent. Of greater importance, these authors, by constructing a model of how people may be hurt in group experiences, conclude that the single most important factor is the skill of the group leader. These authors suggest that 'casualties' occur because 'leaders' put their personal needs ahead of the needs of group members and that they lack empathy. This recent study concludes that the group experience can be used safely to promote personal learning and the development of managers, provided that the potential harm of the technique is understood and controlled.

We believe, although this is not supported by any empirical evidence, that a T-group run by a competent and well-versed leader, can make an important contribution to the personal development *of trainers* about to embark on the design and implementation of social skills programmes. The T-group may have its most important benefits in this role since, on the basis of the research evidence and the costs, it is unlikely to be cost-effective with large numbers of managers.

Role playing   This technique is a particular form of simulation training in which participants are asked to act out prescribed roles in problem situations, often including a conflict of interests of individuals or of the groups they represent. In many respects it is similar to the clinical technique known as psychodrama. The problem, inherent in the role-play exercise, must be solved by the participants, who are given the briefest description of their role and their position on the problem.

Developed by Maier (1953) and his associates, the technique is extensively used in management and leadership training. The objectives claimed to be met by the role-play technique are:

(1) Increased interpersonal sensitivity.
(2) Empathic appreciation of the true position of other persons.
(3) Ability to solve problems 'in real time' using whatever decision-making method seems most suitable, but more often emphasizing the participative strategy.
(4) Providing an opportunity for decision makers to see and 'to talk through' the effect of their decisions on others whose jobs/roles may have been affected.

Like the other simulation methods, the case method and the business game, the strongest criticisms are centred around the problem of artificiality. Complaints of collapsed time frames and insufficient data about the problem are typical responses from trainees to the role-playing experience. The skilful trainer, however, can use the technique to generate valid data about an individual trainee, visible to all, which can be the central focus of the debriefing period. Tragically, however, role-playing exercises are often used merely to fill the space in programme design, and little learning of real value is accomplished. The role-playing technique, to be truly effective, recognizes both trainers' and trainees' consummate skills in being able to give and to receive interpersonal feedback. These are often the very training objectives which are the focus of the training programme.

Hinrichs (1976) lists in addition the following four disadvantages of role playing:

(1) Trainees may feel that the exercise is childish 'play-acting'.

(2) Trainees may revert to overacting and neglect to focus on problem solving.
(3) The trainer has no control over the rewards, which are often in the hands of other trainees.
(4) The technique is limited in the number of people that can be included, is time consuming and is relatively expensive.

Research on the effectiveness of role playing shows that it can be of use in increasing managers' sensitivity to employees' feelings, but an important proviso is that the role-playing exercise must be accompanied by adequate discussion and debriefing (see Lawshe, *et al.* 1959). In addition, as is the case with sensitivity training, the training programme really at best only helps the trainee identify specific behavioural changes that need to be made in the future. No systematic, behavioural shaping and reinforcement is conducted as an integral part of the programme.

We believe that training must be viewed in the context of the organization as a whole. A manager or supervisor may need to learn how to communicate more effectively, but this training should not take place without the participation and commitment of the rest of management, or the group of subordinates with whom he or she has to communicate. Further, if a manager or supervisor is trained in decision-making strategies, does that individual have the power in the organization to implement a decision?

An important new development in the area of role playing, designed to aid transfer to the back-home organization, is the training simulation known as *Looking Glass*. Developed at the Centre for Creative Leadership, North Carolina, *Looking Glass* was supported by a 3-year research grant from the Office of Naval Research. The principal objective was to devise a simulation/role play that would recreate the way in which managers do their work. Specifically, this objective meant that the training device would focus on the leader/manager as part of an organization and would provide the vital organizational context to aid the transfer of training. In practice, the research evidence derived from the running of the simulation supports the notion that managers do *behave* in the simulated environment in a very similar way to the research evidence cited at the beginning of this chapter, i.e. the Mintzberg (1973 and 1975) and Stewart (1967) evidence.

The *Looking Glass* simulation avoids at least the first of Hinrichs' negative aspects of role play and allows the problems of overacting and neglecting problem solving and reward sources to be seen as vital issues of concern to the participants. In addition, perhaps for the first time in management/leadership training, the participants are able to trace, with the help of their trainers, the relationship between the level of social skill competence inherent in the group, and certain end-result 'bottom-line' variables, such as return on capital invested, profitability and opportunity costs.

Finally, *Looking Glass*, as a controlled experimental treatment, allows the comparison cross-culturally of managers from the UK and USA, not only in terms of these bottom-line variables but also in terms of quality of decision making and the ways in which these are related to competence in social skills.

An integrated approach

We suggest that training be viewed in terms of:

(1) Training for a particular job.
(2) Training for a particular company.
(3) Training for life.

Training for a particular job, for example the job of supervisor, would include an understanding of that particular job and the needs as perceived by the person who is occupying that role. Supervisors themselves would provide the best starting point towards this under-standing. Training for a particular company includes all manage-ment levels within the structure of the organization together with their subordinates. Training objectives would include an attempt to achieve greater goal clarity, an examination and sharing of individual and group values and a clarification of roles and functions. Training for life relates to the development of the individual in ways which may not be related directly to the company or particular job. This training emphasizes the continuing nature of the training process and would include the provision of individual or personal development oppor-tunities in relation to needs identified by the person concerned.

This approach to training is based on the following assumptions:

(1) The individual is responsible for his or her own learning.
(2) Individuals differ in their training needs.
(3) Training must include the whole organization.

(4) Training is a continual process, including constant evaluation, practice, feedback and further training as new needs are identified.

A similar approach has been outlined by Thurley and Wirdenius (1973), who believe that there is no basic programme which can be applied to any problem. Instead they identify a need for:

(1) A theory of change and routes to change so that various strategies can be compared.
(2) A comprehensive frame of reference within which appropriate action for particular situations may be planned.
(3) Clarification of possible objectives of change.

These authors propose a model for change which includes steps which they term 'sub-problems' of change. These are:

(1) The analysis and diagnosis of the nature of the situation.
(2) The setting of objectives.
(3) The design of changes.
(4) The evaluation of the results of changes, and the dissemination of such results.

It seems clear that any such approach must include all individuals and groups who will be affected by the changes.

Both approaches outlined above would, we believe, place considerable emphasis on training in the kinds of social skills which we have outlined in previous sections. Further, we would view the same skills as applicable across different management positions. In relation to the managerial roles outlined by Mintzberg we would argue that all managers and supervisors would be expected to spend some time in each of the ten roles he has delineated.

Summary and conclusions

In this chapter we have considered the supervisor as a leader or manager of a working group, and have minimized the distinction between supervisors and higher levels of management. We argue that there are more similarities than differences. We view social skills training as applicable to all levels of management — in fact, the effectiveness of the training might depend on managers at all levels participating including, in many cases, the subordinates.

We identify two major factors which relate to the management of groups:

(1) Group processes and group tasks.
(2) Leadership.

In each case we address some of the relevant research issues, concluding with an identification of a number of group and leadership skills. Our view is that competencies at the group and leadership levels are contingent upon the acquisition of a number of basic interpersonal skills.

Neither supervisory training nor managerial training to date seems adequate for the current needs of organizations. We suggest an integrated approach as a possible future development; this approach would, we believe, necessitate considerable emphasis being placed on the training of social skills.

We believe that the skills which we have identified would lead to more effective management of a working group. Some skills may be more easily acquired than others but, most importantly, they are skills and can be learned; they are not simply characteristics which are either present or absent. Learning effective social skills however, consists of more than learning specific behaviours and then practising them. There is also a set of values, beliefs and attitudes which must be considered in conjunction with the development of a skill. For instance, a person may learn the procedures for consensual decision making but unless that person believes that participation of effective group members will improve the quality of the decision and help its implementation, no real consensual decision making will occur. A skill development programme must focus both on the skills to be learned and the values and attitudes of the learner.

References

Argyris, C. and Schon, D. E. (1974). *Theory in Practice*. San Francisco: Jossey Bass.

Argyris, C. (1976). Leadership, learning and changing the status quo. *In* Hackman, J. R., Lawler, E. E. and Porter, L. W. (eds) (1977). *Perspectives on Behaviour in Organizations*. New York: McGraw-Hill.

Back, K. W. (1957). Influence through social communication. *Basic Studies in Social Psychology*. London: Holt, Rinehart and Winston.

Barber, J. W. (ed.) (1968). *Industrial Training Handbook*. London: Iliffe.

Berenson, B. G. and Mitchell, K. M. (1974). *Confrontation: For Better or Worse*. Amhurst, Mass.: Human Resource Development Press.

Blake, R. R. and Mouton, J. S. (1964). *The Managerial Grid*. Houston: Gulf.

Blau, P. M. (1954). Patterns of Interaction among a group of officials in a government agency. *Hum. Rel.* 7, 337–8.

Blumberg, A. and Golembiewski, R. T. (1976). *Learning and Change in Groups*. Harmondsworth: Penguin.

Burns, T. (1954). The directions of activity and communication in a departmental executive group. *Hum. Rel.* 7, 73–7.

_____ (1957). Management in action. *Operational Res. Q.* 8, 45–60.

Campbell, J. P. and Dunnette, M. D. (1968). Effectiveness of T-group experiences in managerial training and development. *Psychol. Bull.* 70, 73–104.

Campbell, J. P., Dunnette, M. D., Lawler, E. E. and Weick, K. E. (1970). *Managerial Behaviour, Performance and Effectiveness*. New York: McGraw-Hill.

Carkhuff, R. R. (1972). *The Art of Helping*. Amherst, Mass.: Human Resources Development Press.

Culbert, S. A. (1968). The interpersonal process of self-disclosure: It takes two to see one. *In* Hacon, R. (ed.) (1972). *Personal and Organisational Effectiveness*. London: McGraw-Hill.

Dale, E. and Urwick, L. P. (1960). *Staff in Organisation*. New York: McGraw-Hill.

Drucker, P. F. (1954). *The Practice of Management*. New York: Harper and Row.

Egan, G. (1975). *The Skilled Helper*. Monterey, Cal.: Brooks/Cole.

_____ (1976). *Interpersonal Living*. Monterey, Cal.: Brooks/Cole.

_____ (1977). *You and Me: The Skills of Communicating and Relating to Others*. Monterey, Cal.: Brooks/Cole.

Festinger, L., Schachter, S. and Back, K. (1950). The operation of group standards. *In* Proshansky, A. B. and Seidenberg, C. D. (ed.) (1969). *Basic Studies in Social Psychology*. London: Holt, Rinehart and Winston.

Fiedler, F. E. (1967). *A Theory of Leadership Effectiveness*. New York: McGraw-Hill.

_____ (1976). The leadership game: Matching the man to the situation. *In* Hackmann, J. R., Lawler, E. E. and Porter, L. W. (eds) (1977). *Perspectives on Behaviour in Organisations*. New York: McGraw-Hill.

Gassner, S. M., Gold, J. and Snadowsky, A. M. (1965). Changes in the phenomenal field as a result of human relations training. *J. Psychol.* 58, 33–41.

Gibb, J. R. (1974). The Message from Research. *In The 1974 Handbook for Group Facilitators*. La Jolla, Cal.: University Associates.

Graen, G., Alvares, K. and Orris, J. (1970). Contingency model of leadership effectiveness: Antecedent and evidential results. *Psychol. Bull.* 74, 285–96.

Guest, R. H. (1956). Of time and the foreman. *Personnel* 32, 478–86.

Hackman, J. R. (1976). 'Group influences on individuals. *In* Dunnette, M. D. (ed.). *Handbook of Industrial and Organisational Psychology*. Chicago: Rand McNally.

Hackman, J. R., Lawler, E. E. and Porter, L. W. (eds) (1977). *Perspectives on Behaviour in Organizations*. New York: McGraw-Hill.

Hackman, J. R. and Morris, C. G. (1975). Group tasks, group interaction process, and group performance effectiveness: A review and proposed integration. *In* Berkowitz, L. (ed.). *Advances in Experimental Social Psychology*. New York: Academic Press, vol. 8.

Hackman, J. R. and Oldham, G. R. (1976). Motivation through the design of work: Test of a theory. *Organisational Behav. Hum. Perf.* 16, 250–79.

Harrison, R. (1971). Research on human relations training: design and interpretation. *J. Appl. Behav. Sci.* 7, 71–85.

Hinrichs, J. R. (1964). Communication activity of industrial research personnel. *Personnel Psychol.* 17, 193–204.

———— (1976). Personnel Training. *In* Dunnette, M. D. (ed.). *Handbook of Industrial and Organisational Psychology*. Chicago: Rand McNally.

Horne, J. H. and Lupton, T. (1965). The Work Activities of 'Middle Managers: An Exploratory Study. *J. Management Studies* 2, 14–33.

Ivancevich, J. M., Szilaggi, A. D. and Wallace, M. J. (1977). *Organisational Behaviour and Performance*. Santa Monica, Cal.: Goodyear.

Jackson, J. (1965). Structural characteristics of norms. *In* Steiner, I. D. and Fishbein, M. (ed.). *Current Studies in Social Psychology*. New York: Holt Rinehart and Winston.

Jasinski, F. J. (1956). Foreman relationships outside the work group. *Personnel* 33, 130–6.

Johnson, D. W. and Johnson, F. P. (1975). *Joining Together: Group Theory and Group Skills*. New Jersey: Prentice-Hall.

Jourard, S. M. (1964). *The Transparent Self*. Princeton, N.J.: Van Nostrand.

Jourard, S. M. and Lasakow, P. (1958). Some factors in self-disclosure. *J. Abnormal Soc. Psychol.* 56.

Kaplan, R. E., Obert, S. L. and Van Buskirk, W. R. (1980). The etiology of encounter group casualties: 'Second Facts'. *Hum. Rel.* 33, 131–48.

Katz, D. and Kahn, R. (1966). *The Social Psychology of Organisations*. New York: Wiley.

Kelley, J. (1964). The Study of executive behaviour by activity sampling. *Hum. Rel.* 17, 277–87.

King, D. (1964). *Training Within the Organisation*. London: Tavistock.

Kolb, D. A., Rubin, E. M., McIntyre, J. M. (1971). *Organizational Psychology: A Book of Readings*. Englewood Cliffs, N.J.: Prentice-Hall.

Landy, F. J. and Trumbo, D. A. (1980). *Psychology of Work Behaviour*. Homewood, Ill.: Dorsey.

Larson, L. L., Hunt, J. G. and Osborn, R. N. (1975). The great Hi-hi leader behaviour myth: A lesson from Occam's razor. *In* Bedeian, A. G. *et al.* (ed.). *Proceedings of the Annual Meeting of the Academy of Management*. Academy of Management.

Lawler, E. E., Porter, L. W. and Tannenbaum, A. (1968). Managers' attitudes toward interaction episodes. *J. Appl. Psychol.* 52, 432–9.

Lawshe, C. H., Belda, R. A. and Brune, R. L. (1959). Studies in management training evaluation: Chapter 11. The effects of exposures to role playing. *J. Appl. Psychol.* 43, 287–92.

Lieberman, M. A., Yalom, I. D. and Miles, M. B. (1973). *Encounter Groups: First Facts*. New York: Basic Books.

McCall, M. W., Morrison, A. M. and Hannan, R. L. (1978). *Studies of Managerial Work: Results and Methods*. Greensboro, NC: Center for Creative Leadership; Technical Report No. 9.

McGregor, D. (1967). *The Professional Managers*. New York: McGraw-Hill.

Maier, N. R. F. (1953). *Principles of Human Relations: Applications to Management*. New York: Wiley.

March, J. G. and Simon, H. A. (1958). *Organisations*. New York: Wiley.

Mintzberg, H. (1970). Structured observation as a method to study managerial work. *J. Management Studies* 1, 87–104.

_____ (1973). *The Nature of Managerial Work*. New York: Harper and Row.

_____ (1975). The Manager's job: Folklore and fact. *Harvard Bus. Rev.* 53, 49–61.

Ponder, Q. D. (1957). The effective manufacturing foreman. *In* Young, E. (ed.). *Proceedings of the Tenth Annual Meeting of the Industrial Relations Research Association*. pp. 41–54.

Porter, L. W., Lawler, E. E. and Hackman, J. R. (1975). *Behaviour in Organisations*. Tokyo: McGraw-Hill.

Rackham, N., Honey, P. and Colbert, M. (eds) (1971). *Developing Interactive Skills*. Northampton: Wellers Publishing Ltd.

Rosen, N. (1973). *Supervision: A Behavioral View*. Ohio: Grid.

Sayles, L. (1964). *Managerial Behaviour: Administration in Complex Organisations*. New York: McGraw-Hill.

Schein, E. H. (1969). *Process Consultation*. Reading, Mass.: Addison-Wesley.

Sofer, C. (1972). *Organisations in Theory and Practice*. London: Heinemann.

Stewart, R. (1967). *Managers and Their Jobs*. London: Macmillan.

Stodgill, R. M. (1974). *Handbook of Leadership*. New York: Free Press.

Thurley, K. and Wirdenius, H. (1973). *Supervision: A Reappraisal*. London: Heinemann.

Thurley, K. E. and Hamblin, A. C. (1967). *The Basis of Supervisory Training Policy*. Oxford: Pergamon Press.

Timmons, J. A. (1973). *The Entrepreneurial Team: Formation and Development*. Presented at 1973 National Meeting of the Academy of Management.

Vroom, V. H. (1976). Can leaders learn to lead. *In* Hackman, J. R., Lawler, E. E. and Porter, L. W. (eds) (1977). *Perspectives on Behaviour in Organisations*. New York: McGraw-Hill.

Vroom, V. H. and Yetton, P. W. (1973). *Leadership and Decision-making*. University of Pittsburgh Press.

Weissenberg, P. and Kavanagh, M. (1972). The independence of initiating structure and consideration. A review of the evidence. *Personnel Psychol.* 25, 119–30.

6 Presenting and public speaking

CHRISTOPHER K. KNAPPER

Introduction

Presenting as a social skill: scope and definitions

'Presenting behaviour' is a rather clumsy and ambiguous term that is used here to encompass several related social skills. All are concerned with the imparting of information or ideas to transmit knowledge or affect attitudes. Such presentations are a major component of much teaching, especially in higher and further education, where the lecture is a favourite format. Presentations also occur commonly in business and industrial settings, and of course the notion of a 'speech' as a means of communication, motivation, persuasion or entertainment is ubiquitous in western society. Since there is no generally used expression that embraces the educational lecture, the sales manager's pep talk, the Rotary Club after-dinner speech, the sermon and the keynote address to a political rally, the term presentation has been used as the best choice.

Although there are obviously many different ways of communicating information and ideas, the focus in this chapter is on live, primarily verbal presentations by a single speaker to a larger group. Thus presentations through the mass media and informal interpersonal communications are largely ignored. However, certain underlying principles derived from studies in these fields are used to explain aspects of presenting behaviour as defined above. There is one other type of presentation that is technically included within the definition but which will not be discussed in great detail. This involves theatrical performances, such as plays. While these may include a 'public speech' made by one or more individuals to a larger group they are clearly a special case, being relatively 'artificial' simulations of some outside reality and aimed at the immediate entertainment of an audience. Certain notions about effective presentation can be gleaned from a study of theatrical performance and are referred to later, for example in the section on vocal qualities.

Considering that public speaking has been a major form of communication since the beginning of civilization and that the lecture has been a prominent educational tool at least since the Middle Ages ('lektor' was in fact the name given to a religious teacher), the skills used in presenting behaviour are relatively poorly understood, scientifically. One reason for this may be that there is no single effective way of lecturing or public speaking and a great deal depends on the nature of the audience, the subject matter and the situation. At the same time there is an abundance of advice on the topic in popular sources. Most public libraries contain a dozen or so books on giving after-dinner speeches, effective business communication and so on. The topics of 'speech' and 'communication' are the subjects of whole courses in some American high-schools, with the apparent aim of making students orally competent and fluent both at formal (public speaking) and informal (discussion) situations.

Most of the books on public speaking and lecturing supplement the rather scanty scientific findings on the topic with a good deal of anecdotal evidence and common sense, advising their readers to take the fairly obvious steps of planning a presentation beforehand, making sure they speak audibly and so on. While the remainder of this chapter focuses mainly on the empirical research on presentation methods, such common-sense principles are by no means ignored in the advice offered here.

The contribution of psychology

Although the term 'presenting behaviour' would not find a place in a textbook of psychology, many of the principles that can be applied to the social skill of public speaking and lecturing derive from research in a wide number of psychological fields. The most immediately relevant areas are probably communications theory, person perception and social interaction, but research in motivation, learning and memory, attitude formation and change, information theory, emotion and personality has also contributed to an understanding of the presentation task.

Since a major objective of most presentations is to communicate facts, ideas or even emotional tone to an audience, it is not surprising that communications theory has been looked to for insights into ways of improving presentation techniques. Some of the more technical conceptualizations of the communication process, such as Shannon and Weaver's cybernetic model, are too specialized and technical to have much pragmatic value in the understanding of public speaking. On the other hand, mass-communication research has provided a useful general framework that can be readily applied to public speaking.

Lasswell's simple conceptualization of the communication process speaks of a sender, message, communication channel and receiver (Smith, *et al.* 1946).* In more popular terms, Lasswell describes communication as being a question of 'Who says what to whom, by which channel, with what effect?'. The same general framework has also proved useful in person perception, which is concerned with the way impressions of other people are formed (see Warr and Knapper 1968, for an application of the model in personal impression formation).

Applying Lasswell's model to the presentation process, the presenter represents the communication source or sender and is usually the most important determinant of the message. Relevant characteristics of presenters are discussed later, as are qualities of the audience (or receivers) that affect the communication process. The section on presentation modes deals with the question of communication channel: although it may seem self-evident that an oral communication uses simply voice as the channel, the vocal cues are accompanied by a good deal of non-verbal (NV) information from the presenter. Finally, in the second section of the chapter the question of 'with what effect?' is addressed in a discussion of the goals of presentations, especially the distinction between communicating information or ideas and changing attitudes.

Presentation goals

Presentations differ widely in the ends they hope to achieve. In the lecture the aim is usually to impart information with a view to increasing knowledge or understanding. Other presentations are

---

* This general model did not originate with Lasswell. Aristotle in his *Rhetoric* wrote that three elements basic to any communication are a speaker, a message and a listener.

concerned with teaching skills, ranging from salesmanship to chess. In still other cases the speaker is primarily interested in persuading the audience to a particular point of view, arousing them emotionally, or even simply enhancing the presenter's image. Of course many (perhaps all) communications succeed in doing a little of all these things. However, the presentation techniques that are most effective for one purpose are often much less effective for another. For this reason, and as a general organizational principle, it is wise for presenters to decide in advance upon their objectives.

Imparting information

Perhaps the most common use of the set speech is to convey knowledge to the listeners. This is especially true for the academic lecture, which has been the subject of a good deal of research aimed at investigating its effectiveness in comparison with other teaching methods. Bligh (1972) summarized the results of 107 studies that examined the amount of learning resulting from a lecture compared with methods such as discussion groups, seminars and so on. The tests of learning varied from study to study, but most commonly included a multiple-choice retention test given shortly after the end of the teaching session. In twenty-four of the studies students in the lectures fared better, in twenty-one cases lectures produced poorer learning, and the remaining sixty-two comparisons found no significant differences between the methods used. On the basis of these results Bligh concluded that lecturing can be as effective as any other method of transmitting information but that it is not demonstrably superior.

Attention and memory The aim of most teaching is to facilitate transfer of knowledge to new situations outside the classroom. This requires a type of learning that goes beyond the mere acquisition of facts to a deeper conceptual understanding, and ultimately the application of basic principles to real life. The lecture's ability to achieve such ends is discussed in the next section.

Meanwhile, however, it is reasonably obvious that facts that are simply not acquired in the short term cannot be built upon later, and hence there is interest in knowing in what circumstances retention of information communicated in a formal presentation can be maximized. In other words, what can the presenter do to prevent what is said from being rapidly forgotten? Investigations of human

memory show that a good deal of apparently 'forgotten' material ha
in fact never been registered by the listener, even though it woul
have been heard (Hunter 1964). Thus an important principle o
communication is that the message must be attended to.

There are various ways in which a presenter may try to increase th
attention paid to what is being said, and which the psychologica
literature suggests might be effective. For instance, in introducin
facts or principles it is preferable to follow the general statement b
specific examples. Another useful strategy is to introduce variety int
the presentation, perhaps by use of audio-visual aids, demonstrations
or even inserting small-group discussion exercises. If this is no
feasible then the speaker should imitate the good actor, and at leas
attempt to vary vocal presentation as much as possible. A furthe
technique that the learner may use to focus attention is notetaking
although this may also have some disadvantages, which are discusse
in the following paragraph.

It is well established by research — as well as common sense — tha
information is not generally remembered without the opportunity t
become 'embedded'. This usually implies some further processing o
the material (e.g. revising later from lecture notes). Miller (1951) ha
argued that the 'short-term memory' is normally incapable of storin
more than seven pieces of information. Miller was talking in th
context of memory for fairly simple types of material, such as words o
numbers, and hence it is not clear how much the often quoted figur
seven is applicable to the lecture or speech. Here the informatio
being presented usually encompasses a mixture of simple facts an
more abstruse generalizations, and the difficulty of analysing
presentation in terms of the amount and type of information trans
mitted makes it virtually impossible to compare one lecture or speec
with another. None the less, it is clear from investigations of memor
for lecture material that a very great deal is lost or forgotten for on
reason or another (see McLeish 1976). Apart from lack of attentio
other factors that help to induce forgetting are 'interference' fro
other lectures, books, etc., or indeed between different parts of th
same presentation. Even notetaking, which may facilitate studen
attention, may interfere with listening. The situation is complicate
further when the information or ideas are badly organized, confusin
or presented too rapidly.

What does this mean from a practical point of view? First, variet
can and should be introduced into the presentation in the way

discussed above. Second, presenters should be aware that only a limited amount of material can be absorbed by most of their audience during a fixed time period. Perhaps the most common fault of lecturers is to include far too much material, on the mistaken assumption that everything transmitted is received. Even the amount of time allotted for the presentation may militate against effective communication: for example, MacManaway (1970) argued that 20–30 minutes is the optimum time for a lecture, after which attention (and learning) falls off rapidly. Third, presenters could allow their audience to receive and process the information by building a certain amount of redundancy into their remarks and by repeating important points. Fourth, presenters should strive for clarity in both presentation method and organization of material. Avoiding confusion on the part of students is easier said than done, but one way of aiding the audience in this respect is to try to make the information as meaningful as possible by use of suitable examples and by relating concepts to their own experience. The more technical aspects of organizing concepts for presentation are touched upon in the next paragraph. Fifth, notetaking may help students to focus attention and may aid long-term retention of information.

Organizational structure   So far little has been said about the internal structure of a presentation: the different ways in which the information and ideas can be organized for effective communication to an audience. While the organization of ideas may reasonably be as important as the way they are presented (in terms of number, vocal quality, etc.), the most effective mode of structuring a lecture will probably vary considerably according to the particular circumstances in which it is delivered. The difficulty of the subject matter and the level of comprehension and ability of the audience seem likely to be especially important factors.

As long ago as 1910 John Dewey suggested a problem-based approach to presentations, and he outlined several basic components, including an attention-getting introduction, a statement of the problem, possible solutions to the problem, and an appeal for the audience to implement the best solution. A great deal could be said about how to structure the individual components and the interested reader is referred to Wiseman and Barker (1967) for an account of how this may be done. Not all lectures, of course, are problem centred. Bligh (1972) reviews the most common forms of organizations, including 'classification hierarchies' (grouping different points of information together with some kind of unifying feature as a heading), 'chaining' (a sequence of ideas or events told as a narrative), a 'thesis' (a combination of chaining and problem-centred forms), etc. Since the organization of material for presentation is a specialized matter, the reader is referred to Bligh (1972, Chapter 3) for further details.

Empirical research to date cannot provide definitive answers about the long-term effectiveness of different presentation methods. In the educational system especially, exposure to information in a lecture is generally only one source of possible information for the student, and hence any ultimate learning gain may be attributable to numerous other factors, such as the amount of independent reading.

Teaching concepts and skills

Although the lecture method may be reasonably successful for imparting basic information, the presenter, particularly in educational settings, often has more lofty aims in mind and intends the audience to go beyond the mere acquisition of facts to some more basic understanding of underlying concepts and theory. In still other circumstances the aim of the presentation may be to teach some sort of skill, either verbal or physical. As some of the remarks in the preceding section may have shown, the 'straight lecture' is not a particularly good medium for achieving these goals.

The number of empirical studies comparing lectures and other techniques for their ability to achieve concept learning is considerably less than for basic knowledge acquisition. This is probably because the dependent variable of conceptual thinking is relatively difficult to measure. Of the reports that exist, however, neither Bligh (1972) nor Dubin and Taveggia (1968) were able to find a single one where the lecture was superior to other methods (usually discussion techniques) for the 'promotion of thought' or problem solving. Earlier Bloom (1953) had students record their thoughts during taped lectures or discussions and found that the former provoked little synthesis or problem solving and a high proportion of passive or irrelevant thoughts. One reason for this, it was suggested, is that students in many educational settings go into a lecture with the expectation that they will not have to do much in the way of processing information but can simply sit back and let it flow into (or over) them. Psychological studies of learning certainly indicate that for students to become adept at any skill, including problem solving, there must be some practice or activity, which is not usually so in a lecture. Simply knowing the principles of something is not the same as being able to apply them. In manual skills the importance of activity and practice is even greater: for example, it would clearly be absurd to expect anyone to learn how to play the violin merely from listening to a lecture on the subject, however authoritative the presenter, and however inventive the presentation techniques.

Assuming that problem solving is the aim of the presentation and that activity at the task is not possible during the session for logistical reasons, then at least the lecture may be arranged so as to stimulate further exploration once the presentation has ended. This may be achieved by assigning specific tasks or exercises, or by adopting an 'open-ended' approach to the presentation, in which certain issues are left unresolved. Several investigators have concluded that complete and precise organization of a presentation may have negative effects on later learning, perhaps because students see such issues dealt with in such a lecture as 'cut and dried' and in no need of further thought (Petrie 1963; Thompson 1967).

Changing attitudes and increasing motivation

In academic presentations the goal of changing attitudes or enhancing student motivation is generally much less important than imparting information or ideas. The few studies that have looked at this question show that lectures do little to change student attitudes to the discipline or to stimulate motivation for further study of the topics covered (see Hartley and Cameron 1967). Outside academic presentations there are numerous situations where the main goal of a presentation is to change attitudes or enhance motivation, and the ability of different presentation styles to do so has been the subject of considerable interest and research, a good deal of it in the field of persuasion and propaganda. In the 1950s Carl Hovland and his colleagues conducted a series of famous studies at Yale University to investigate techniques of effective oral propaganda (see Hovland, *et al.* 1953). Their research subjects comprised both students and members of the American Army.

Hovland and his associates were able to identify a number of factors that contributed to attitude change on the part of listeners. In the first place the *credibility* of the speaker was of great importance in persuading the audience of a particular point of view — a finding that has often emerged in other contexts, notably person perception (see Warr and Knapper 1968). Credibility here seems to be equated with 'authority' on the part of the speaker, which can either derive from some previous status and reputation (a title or simply a good deal of relevant experience) or can be based upon performance during the presentation itself. It seems to be preferable if the presenter does not boast of his own achievements but allows his manner to communicate confidence and knowledgeability (Argyle 1978). Some commentators add, however, that the aura of authority should not be exaggerated so much that an error or lack of knowledge may completely undermine the speaker's credibility: hence presenters are advised to admit ignorance of certain topics when necessary.

The second general finding of the Hovland group was that, contrary to some expectations at the time, communications based upon the *arousal of fear* in the audience were not particularly successful. Later research on 'cognitive dissonance' (see Festinger 1957) has explained this in terms of the audience's simply discounting information that is too frightening for them to cope with effectively. Other research has shown that a general level of arousal or emotional excitement may be more successful in inducing attitude change, especially if followed by some definite solution or course of action (Brown 1963; Sargant 1957). Observation of presentations by evangelists like Billy Graham tend to confirm the validity of such an analysis.

When a presentation is primarily intended to persuade, then it is common to select the information so that the speaker's main argument is placed in the best light. On the other hand, there is always the chance that this will rebound and cause members of the audience to lose faith in the speaker when they are later exposed to omitted material and alternative viewpoints. Hovland concluded that *giving only one side of an argument* could be effective in some circumstances but that giving both sides of the issue was a preferable tactic if members of the audience were intelligent, were likely to be in disagreement with the speaker and were at some time likely to hear the other side of the question.

A related question investigated by Hovland and numerous others is the best order to present information. Hovland's group was particularly concerned with presentation order for chunks of material espousing different points of view about an issue. Although not all studies support the conclusion, the general weight of evidence suggests a 'primacy effect', meaning that it is best to present first the favoured side of the argument. This seems to be because the initial material has a large effect on the impression of the audience, and they have a tendency to interpret later information in the light of this first impression. The same appears to be true not only in responding to persuasive arguments but in forming impressions of other people (Warr and Knapper 1968).

A most important question is how attitudes relate to behaviour. It is one thing to persuade someone a particular course of action is desirable, and quite another to get that same person to behave in accordance with beliefs. The question of how best to affect behaviour by persuasion was investigated during the Second World War by Lewin in a project concerned with how to convince people to change their eating habits during food shortages. Lewin (1943) compared different presentation methods and concluded that having people make a group decision on the basis of discussion was considerably more effective for changing their behaviour than providing them with a lecture. The effect could well be due to the sorts of group conformity pressures that exist within such a discussion group.

The above discussions of speaker credibility and order effects show that what happens at the beginning of (or even before) a presentation may colour the impression the audience forms of its content. This phenomenon is known as *mental set* and has been widely observed by psychologists in a number of situations. A dramatic demonstration of set in the classroom was provided by Kelley (1950), who had students listen to a guest lecture after receiving some initial information about the lecturer's qualifications and personality. This information was identical for all students except that half the group were told that the speaker they were about to hear was a 'warm' individual, while the other half were told he was a rather 'cold' person. This manipulation had a dramatic effect on students' perceptions of the speaker, even on traits that would appear to be completely unrelated to the dimension of warmth (such as intelligence). Further, students in the 'warm' group displayed significantly more interaction with the speaker in a discussion period after his presentation. Although Kelley did not measure the amount of learning from the lecture or belief in the speaker's arguments, it seems plausible that there might be a difference for the two groups largely dependent upon the initial, and arbitrary, impression that they had formed.

In this case the quality of warmth attributed to a speaker seemed to have certain positive effects on impression formation and interaction. This probably relates to the rather mystical concept of 'rapport' that is thought to be an important quality in such diverse areas as the theatre and psychotherapy as well as in public speaking. Certainly a warm and caring attitude on the part of university lecturers is believed by many students to be an important quality that contributes to teacher effectiveness (see Sheffield 1974). At the same time the evidence linking such qualities as warmth with learning effectiveness is much more equivocal, and listeners may easily have favourable attitudes to a speaker and yet learn nothing. A dramatic illustration of this was the so-called 'Dr Fox effect', in which a professional actor gave a polished and entertaining lecture whose content was non-sensical. Yet student ratings about the 'learning experience' were extremely favourable! (Naftulin, *et al.* 1973.)

Although rapport is still an ephemeral concept that remains largely uninvestigated outside therapeutic fields, most people would agree that such a phenomenon exists, even if it is difficult to define. One way of achieving rapport may well be for speakers to be sensitive to audience needs, or to try to establish some sort of common bond or identity with listeners, provided that this does not interfere with the sense of authority that makes the speaker worth listening to in the first place. Many popular commentators on public speaking mention humour as a way of achieving these ends, especially humour at the speaker's own expense. There are few empirical data to link the use of such humour with audience attention or learning. At the same time there is evidence that many different speaking styles can be effective as communication devices and that a humorous presentation may be effective in some circumstances but not in others — for example, when the hilarity is forced. It may also transpire that although a humorous approach has a favourable effect on attitudes, it fares less well for learning: in other words, listeners remember all the jokes but not the information it was hoped to convey.

Presentation modes

'Modes' of presentation simply means the primary method used by the speaker to communicate with an audience. At the beginning of this chapter discussion was limited to oral presentations, thus excluding the many alternative ways of providing information and ideas, ranging from the use of models to slide—tape displays. Even oral presentations, however, turn out to be quite complex; further, although the voice may be intended as the primary communication channel, information is being transmitted simultaneously in other ways to provide a great many NV cues to the audience. The present section discusses both verbal and non-verbal information and ends with a discussion of how different presentation methods may be combined for maximum effectiveness.

Verbal methods

Vocal quality   Since giving a speech by definition includes primary reliance on the voice, it is not surprising that a great deal of attention has been given in books on public speaking to using the voice as effectively as possible. This appears to include audibility — especially avoiding the common fault of dropping the voice at the end of a sentence — intonation (which includes loudness, speed of delivery, variety of pitch, accent, etc.), and fluency: avoiding too many 'ers', long pauses or periods of silence. Training in vocal technique is an integral part of the curriculum at drama schools and is sometimes used to help university and college lecturers. Murray and Lawrence (1979) have demonstrated that a course of this kind can have a beneficial effect on student ratings of instruction.

All the qualities mentioned above are thought to relate to comprehensibility of a presentation. In addition, a speaker's accent can also be a cue to certain personal qualities, which may affect credibility. For example, it is commonly believed that the BBC uses news readers who have a standard 'Oxford English' accent to lend credence to the material they are presenting. On the other hand, the same accent may be less effective in the case of a blue-collar trade union leader addressing a group of Cockney dockers. Effects of a speaker's accent on audience perception have been demonstrated in a number of empirical studies (e.g. Giles and Powesland 1975; Scherer 1971).

Other qualities of voice can communicate emotions such as optimism or anger — as is often seen in political or religious orations. One way of dominating an audience is by means of a rapid, loud and constant barrage of words that forestall any interruption.

Non-verbal modes

As mentioned above, the speaker who believes his verbal message to be the only one reaching the audience is naïve indeed. Not only can many of the subtle vocal qualities described above affect perception of a communication, but in addition many aspects of a speaker's appearance and physical behaviour can be transmitted with the oral message, and these too can change the audience's reaction to what is said.

Appearance Public speaking, like all social skills, includes an element of role playing, acting or what Goffman (1959) calls 'impression management'. This sense of 'presence', which is characteristic of a successful performing artist, seems to exist in the case of effective propagandists or evangelists. The qualities that make a speaker appear to be 'larger than life' are not solely visual and may include some of the vocal qualities mentioned above (Hitler would be an infamous example).

At the same time, there is a good deal of research evidence that various aspects of physical appearance can also have a marked effect. A good example here is the way a person dresses. Clothes (as well as such accessories as lipstick and spectacles) have been shown to relate significantly to judgements about the wearer's personality. For example, the same individual dressed differently will often be judged to have quite different personal traits, especially when perceivers have not much other information to go on (see Ryan 1966).

Although there has been virtually no research on the effects of clothing on perception of a speaker's message, since clothing has been shown to relate to judgements of liking and esteem it seems reasonable to assume that such judgements may spill over into attitudes towards what the wearer of the clothing is saying. If, as seems likely, clothing does influence the effectiveness of public presentations then it probably does so in the same ways as a speaker's accent, helping to establish a mental set that may colour interpretations of what is said. Thus the way the process works will depend as much upon the nature of the audience as on the presenter.

Incidentally, there is evidence that judgements about personality made on the basis of clothing style are often irrational. Knapper (1970), for example, showed that there were no systematic relationships between perceived clothing style and the 'real' personality of the wearers, as determined by scores on a personality test.

Other aspects of a speaker's appearance include body orientation or position in relation to the audience. For example, should the presenter stand on a rostrum or behind a lectern, or would it be better to sit in a circle with other members of the group? This will partly depend upon the size of the audience, but sitting as part of a circle may aid interpersonal communication by facilitating eye contact between speaker and listeners. Use of a rostrum or lectern may serve a purpose in some respects (ease of consulting notes or increased visibility to the audience); on the other hand these devices may serve as 'barriers' between speaker and listeners, increasing the psychological as well as physical distance between them. There is little empirical evidence on whether bodily posture is an important factor in communication effectiveness, but common sense would suggest that a speaker who keeps eyes glued to notes and body hunched away from the audience will be less effective than a lecturer who often looks up and adopts a physical stance facing the listeners − all other factors being equal.

Many presenters have habitual gestures that may, if repeated often enough, become so irritating that they interfere with the communication process. Habits such as head scratching, swaying or pacing up and down are often mentioned negatively on student rating forms for lecturers, along with various vocal 'tics' such as the repeated use of a favourite phrase. Some movement during an essentially verbal presentation appears to be inevitable and may serve a useful purpose by adding emphasis to points made orally. Dittmann (1972) found that there is a general bodily movement during speech which he attributed to a kind of energy overflow. One way of correcting an inappropriate use of such extra energy is to record a rehearsal of the presentation on videotape (or even in front of a mirror) so as to become aware of how the speaker looks to an audience.

Facial expressions and eye contact   Of all the areas of the body that can transmit information to another person, none is more important than the face. There is good evidence that the facial area receives by far the greatest amount of attention when one person looks at another (Argyle 1975). Although the permanent features of the face (long noses, high foreheads, etc.) are quite unreliable cues to the possessor's underlying traits, 'temporary' facial expressions (smiles, frowns, etc.) have been shown to be both valid and reliable indicators of mood and emotional state (see Argyle 1975; Secord 1958). Even such subtle cues as the extent of pupil dilation can be used to assess mood with some accuracy, although it is unlikely that this would be possible in most public speaking. A well-known trick in preparing a choir for competition (which I have successfully used) is to have the singers smile constantly during the performance: this is thought to convince the adjudicator that not only are the singers happy but also that the music itself is 'brighter'. Assuming certain expressions for effect may sometimes backfire, however, for there is evidence that many observers are able to detect the difference between a false expression and a true one.

Probably of more relevance to public speaking is the importance of maintaining eye contact between presenter and audience. Looking at the listeners enables a speaker to obtain feedback about the reception a message is having. Here eye contact, in the strict sense, may not always be necessary, since a certain amount of information can be obtained simply by seeing whether listeners are facing the speaker, making notes, talking and so on. Eye contact goes further by facilitating a certain degree of intimacy between speaker and listener. Exchange of glances also provides a source of reward for the parties affected especially when these are accompanied by nods, smiles or prolonged gaze. Finally, eye-glance exchange serves the additional purpose of promoting interaction: for example, looking back toward a listener after making a long speech signifies that it is his or her turn to make a contribution. Of course this particular cue is not very important in a one-way presentation where no discussion is permitted.

Eye contact can sometimes have disadvantages. Argyle and Cook (1976) have described how being the subject of visual scrutiny can cause high levels of anxiety in some speakers, especially when they are novices. One way of ameliorating such nervousness is to reduce the level of visual contact between speaker and audience — for example, by speaking in the dark (presumably with the aid of slide projector!) or by making the presentation from the back of the room. Such unorthodox approaches to public speaking seem unlikely to win many friends or influence audiences, especially since they would appear to act against the presenter's chances of improving performance, which will greatly depend on the amount of feedback it is possible to obtain. Certainly for most public speakers and lecturers a sound piece of advice is to maintain frequent eye contact with as many members of the audience as possible, by means of 'sweeping' the listeners with a glance from time to time.

Choice and combination of presentation methods

Since the term 'presentation' as used here is restricted to live oral lectures or speeches, consideration of such additional communication methods as audio-visual aids, discussion-group techniques and so on is, strictly speaking, out of place. They are mentioned, however, to draw attention to the fact that in many situations a presentation that is basically vocal can be embellished by using techniques that may be more suitable for some purposes, and hence more effective. Thus the use of a slide, model, film, audio-recording, overhead-projector transparency, or simply a well-drawn diagram on the chalkboard may illustrate a point that is difficult to make orally as well as introducing the necessary variety that has been stressed earlier. The precise way this may be achieved most successfully is clearly not a suitable topic for the present discussion. McKeachie (1978) provides a good overview of the different presentation modes available.

Although a lecturer is often constrained by logistical factors (e.g. the necessity of communicating with a large audience within a short space of time), there are many circumstances where the one-way oral presentation can be supplemented by interactive techniques that are more flexible and permit a greater degree of activity and practice by the audience. Bligh (1972) and McKeachie (1978) have both shown how small-discussion groups of three or four people ('buzz groups') can be formed and made to work effectively even in the context of a large, relatively formal lecture. The same authors also suggest other discussion-based techniques for rendering the lecture more effective.

Another supplementary device is that of student notetaking, which is clearly out of the direct control of the presenter but which at least the lecturer may encourage and guide. Howe and Godfrey (1977) have summarized the research evidence on the usefulness of student notetaking, and it seems to be the case that the practice not only helps later recall of information but can also perform an integrative function during the lecture itself by helping the student impose some organization on the material being presented. The lecturer may even be able to aid this process by providing a lecture outline, preferably comprising a series of open-ended questions to be addressed by the listener, rather than simply a list of definitive points to be made.

The preceding discussion shows that there are many circumstances in which an orthodox lecture is not the best way of communicating with an audience, and the potential lecturer or public speaker would do well to consider very carefully the alternative presentation methods that may be available, some of which were discussed above. The formal lecture is probably often used in teaching situations almost 'by default' because it is a familiar method to instructors and an expected method for students. In view of the somewhat pessimistic findings about the effectiveness of lecturing for many purposes, a useful starting point for the public speaker in planning a presentation might be to ask in what way the lecture will be superior to, say, simply giving a printed version of the lecture notes. As long ago as 1928, Greene showed that reading a passage produces just as much learning as listening to a lecture on the same topic.

This is not to say that public addresses cannot be extremely effective. The fact that large numbers of people will often turn out to hear them seems to attest to the inherent attraction of watching a live performer — especially a famous one — when an increasing amount of communication comes through the electronic media. The lecture or public speech, then, is often a good way of reaching large numbers of people face to face. It combines a relatively rigid organizational structure (because it is under almost exclusive presenter control) and the potential of some flexibility, provided that the presenter is able to respond to NV cues from the audience. In this way a sensitive lecturer can adjust the presentation in terms of pacing, content and organization of material so as to reflect student knowledge and needs. This is not usually possible for communication through the media (although some steps are being taken in this direction with the introduction of interactive radio and television).

Another potential advantage of the oral presentation is that it can contain very up-to-date information; the only limitation here is the awareness and energy of the speaker.

Limiting factors

Preceding sections have already discussed a number of factors that may limit the effectiveness of a lecture presentation. For example, the lecture lacks two basic essentials of learning: first, that students should participate in the learning process; and second that they receive some sort of feedback on their progress (either from their teacher or on the basis of their own performance). It is because of this sort of limitation in the lecture method that presenters are urged to make a careful assessment of their aims when preparing a lecture and to consider combining the traditional oral method with some of the other approaches described above. Although some limiting factors are partly within the control of the speaker, there are other matters which lie almost entirely outside the presenter's influence. These include the characteristics of the audience being addressed and the situation in which the presentation is taking place.

Audience characteristics

Individual differences   Many audiences have characteristics in common. For example, university students comprise a fairly homogeneous population in terms of age, ability and even attitudes; attenders at a service-club luncheon will also probably be of roughly equal socioeconomic status and may share a good many basic values. But psychologically the homogeneity is rather superficial and disguises many subtle differences in learning and thinking patterns between individuals. This means that what is an effective communication for one person may be less effective − or quite ineffective − for another. Hence it is not surprising that the early students of communication (e.g. Hovland, *et al.* 1953) and later researchers in person perception (e.g. Warr and Knapper 1968) identified differences between 'receivers' as a major variable in their investigations. In the more general area of teaching and learning there is an increasing recognition of the importance of matching instructional methods with different learning styles (e.g. Hunt and Sullivan 1974; Perry 1970).

The lecture, by its very nature, is generally unable to take account of such differences between members of the audience being addressed. However, a public speaker can certainly remember some of the basic characteristics of listeners in both the preparation and delivery of the presentation. In an academic setting a major factor is obviously the ability and knowledge level of students, which might be ascertained by an informal question or quiz early on in the address. In public speaking cultural factors and social expectations of the audience may affect reception both of the speaker and the message. For example, for a lecturer to wear an academic gown at Oxford University would be commonplace, whereas to do so in most North-American universities would be regarded as bizarre. Similarly, the use of *risqué* jokes might endear a speaker to an audience of businessmen, but be perceived very differently by a class of feminists. Perhaps more importantly, the whole 'style' of a presentation may be more or less effective for different cultural groups. Some of the discussion-group techniques recommended above may, for example, be ineffective in very formal situations where participants are unwilling to participate during the course of a presentation. (Argyle (1978) mentions that the use of T-groups has been much less successful in Britain than in North America, largely because of unwillingness by some British participants to 'play this game'.) Even those aspects of interpersonal communication that might be thought to be relatively 'fixed', such as facial expression or patterns of eye contact, can sometimes communicate different things in different cultures (Argyle 1975).

Motivation   Listeners will pay more attention and learn more if they are highly motivated. Unfortunately, psychologists are less certain about how such interest may be stimulated. One obvious step the presenter may take, however, is to ensure that the address goes some way to meeting the goals and needs of the audience: this might be confirmed by some initial probing questions. For academic present-ations, some university teachers may object to this advice on the grounds that they have material 'to cover' in a course or lecture, regardless of the interests or aptitudes of the students in the class. However, as Bligh (1972) points out, there seems little point in conscientiously covering material that is simply not comprehended by most of the audience. While no lecturer can guarantee student interest in all the areas to be taught, student motivation is something that is dangerous to ignore in all but the most autocratic educational settings.

Establishing the appropriate mental set at the beginning of the presentation is also helpful in encouraging audience interest and motivation. This has sometimes been done in dramatic ways: for example, one psychology lecturer in a North-American university is well known for periodically appearing in the guise of one of the great men whose work is being discussed. A more modest way of establish-ing a favourable set at the start of a lecture might be to introduce an interesting problem or question that is to be answered in the course of the presentation. If the question is something of a paradox that makes it additionally intriguing, then so much the better (Katona 1940; McKeachie 1965). Even here, however, the role of individual differ-ences between students may be important. For example, Feather (1969) showed that some students resist this type of challenge and prefer lectures that are quite predictable.

In presentations that are intended to persuade rather than inform, there seems to be some evidence that attitudes are more likely to change under conditions of high emotional arousal (Sargant 1957), and effective propagandists are often very successful at establishing such an emotional tone early on in their presentations.

The presentation context

The situation in which a presentation is made may often be outside the control of the speaker; none the less, contextual variables most certainly affect the reception of what is presented. One such variable is the *size of the audience* being addressed, and class size has been the topic of several investigations of student learning in academic settings. There are conflicting results from these investigations, but McKeachie (1978) reports that the weight of the evidence seems to support the contention that teaching smaller groups (up to twenty students) is more successful for information retention, problem solving and the development of sophisticated attitudes within the discipline. Size of group may be less important in the case of the simple transmission of knowledge – which is what lectures are best equipped to do. Although a speaker may not be able to control the size of the audience, it may be feasible to modify the presentation techniques to deal with smaller or larger groups of listeners. Obtaining feedback by eye contact is relatively simple if the group is small, but impossible for an audience of several hundred. Even for very large groups, however, it is possible to get feedback; for example, by use of the buzz-group technique described earlier.

Other situational factors that may affect communication are the *length of time* allotted for the address, the *time of day* and even the *atmosphere* or *decor* in the room (see Bligh 1972). For an unduly long presentation time the lecturer can at least arrange for breaks to be taken; but the other factors will usually be beyond direct control.

Ways of improving presentations

Are good speakers born or made?

To anyone who has read this far, it should be obvious that there is no such thing as an 'ideal' presentation style or format. Nevertheless, the notion that some individuals are 'born speakers' and that the art of public speaking is some kind of mystical process that cannot easily be learned is still very common. There are certainly innate qualities that may be helpful, such as a mellifluous voice or an imposing stature. On the other hand, many of the techniques of effective presentation can be acquired through study and practice.

Because of the many different presentation contexts, the different goals of public speaking and the influence of different audience characteristics, there is no single style of presentation that can be recommended. It is for this reason that Sheffield in his study of teaching in the universities gave his report the title 'No one way' — meaning that the lecturers identified as effective by his sample of former students appeared to adopt very different presentation methods (Sheffield 1974). Even Sheffield, however, agreed that there were some common characteristics among the traits ascribed to effective university teachers, and this finding has been confirmed in many hundreds of analyses of student ratings of instruction. Good teachers (at least as perceived by their students) are usually those who have a solid mastery of their subject matter, are capable of organizing material effectively and adding appropriate emphasis to the major points, display warmth and enthusiasm and are reasonably flexible in their approach to teaching (see Eble 1976).

Some training techniques

Assuming that effective speaking can be taught, then what techniques are most effective? Since public speaking is a complex skill, *practice* is essential for perfection. Practice, however, should be combined with some form of *feedback* to the presenter. For instance, a speaker might be trained to respond to the various NV cues that emanate from the audience and which can communicate a great deal about their reactions (e.g. yawns, lack of eye contact, notetaking behaviour and so on). Argyle (1978) showed that people can indeed be trained to be more sensitive to facial expression and NV cues when speaking. If the situation is sufficiently informal the presenter can also obtain direct verbal feedback by probing members of the audience to determine their level of comprehension or mood: however, to obtain honest responses may require considerable effort and skill. Yet another form of feedback is to rehearse in front of an observer. This could take the form of a colleague or a mechanical recording device.

The idea of videotaping presentations for later review by the speaker and a professional consultant has grown in popularity in recent years and has been shown to be highly successful if managed properly. Without the presence of a skilled consultant, however, the technique may provoke more anxiety than improvement.

Another social learning technique that is well known to psychologists is *imitation* or *modelling*. One problem here is that, as has been argued above, no single model exists for effective public speaking. On the other hand, observation of the salient behaviours displayed by successful presenters can be used as a basis for improving one's own performance. Indeed this probably happens anyway, as it does in so much other social behaviour. Certainly in the case of vocal training, the basic technique appears to be demonstration by an expert teacher, followed by practice by the trainee presenter, and Murray and Lawrence's (1979) study gives some support to the idea that this can be a successful learning strategy.

One special problem with public speaking, as for all public performances, is that of *stage fright*. Studies of the effects of arousal level on performance show that there is an optimum level of emotional arousal in most tasks: in other words, a certain amount of stage fright (as many actors will testify) seems to stimulate a good performance, but too much can be debilitating. In public speaking, arousal level is likely to be affected by such factors as the experience of the presenter, the importance of the occasion and the size of the audience.

Anxiety is often greater before beginning a presentation ('waiting in the wings') than while actually speaking, when the task itself requires full attention. Of course if the speaking task is handled badly then nervousness may well increase and the ultimate result can be a 'vicious circle' of paralysing hesitancy and stage fright. One obvious remedy in both circumstances is for the speaker to boost self-confidence by having prepared in advance a careful plan for the content and format of the presentation. Where the environment is unfamiliar, the speaker might also check such matters as the properties of the room, availability of necessary visual aids, probable audience size, etc.

For many beginning speakers a particular source of nervousness relates to being the centre of attention from a large, impersonal group. If so then this is yet another reason for steering clear of the 'straight lecture' and introducing some of the other presentation methods discussed earlier in the chapter, such as discussion techniques or use of audio-visual aids, which will require less of a 'performance' by the presenter and place more emphasis on diverse sources of information. As mentioned above, the novice speaker who is especially nervous may be able to reduce anxiety by lessening the amount of contact with the audience. On the other hand, this would appear to be a poor long-term strategy in that it cuts off feedback cues that are important in learning to make effective presentations. Probably the best solution for most public speakers is simply to practise speaking as much as possible, beginning with non-threatening situations (before groups of friends or in other informal settings) to become more confident in larger and less familiar situations. Such 'desensitization' techniques have been used successfully in therapeutic settings to help chronically nervous speakers.

Some practical hints

One truism about the social skill of public speaking is that it cannot be learned or improved simply by reading this (or any other) chapter. Nevertheless, some of the comments above may be used as a basis for future activity aimed at perfecting presentation skills. The following is a list of the most important practical principles:

(1) Keep in mind that there is no single effective way of making an oral presentation: a great deal depends upon the audience, type of subject matter, and the situation.

(2) Begin by setting out objectives for the presentation well in advance: in particular decide whether the main aim is to impart facts, teach concepts or skills, change attitudes or some combination of these.

(3) If the main aim is to achieve conceptual thinking, teach skill performance or influence attitudes then consider some alternative method to the straight lecture.

(4) Decide whether you are interested in an immediate response from the audience, or whether you are more concerned about long-term goals (learning or changes in attitude).

(5) Be prepared for the presentation: this not only includes being knowledgeable about the subject matter, but also knowing well in advance how the material is to be organized, and what presentation techniques are to be used during the address.

(6) To help your audience follow the presentation, consider

providing an outline (verbal or written) at the start of your remarks and a summary at the end.

(7) Introduce attention-getting focusers to emphasize important points in your presentation: these may take the form of headings noted on the blackboard or overhead-projector transparency.

(8) Introduce variety into the presentation by means of audio-visual aids (slides, overheads, demonstrations), or even break for small group discussion from time to time (buzz groups).

(9) If audio-visual aids are used, make sure that they are appropriate (i.e. that slides are used to present primarily visual information, the blackboard is used only for diagrams and expressions that are difficult to transmit orally); make sure that they are well prepared (e.g. can be read clearly from the back of the room).

(10) Speak loudly and clearly (this can be checked in advance by having a colleague listen from the back of the auditorium) and introduce variety into your vocal presentation by changing intonation; in particular try not to drop your voice at the end of each sentence.

(11) Allow time for listeners to process what is being said: introduce a certain amount of repetition of key ideas and redundancy (expressing ideas in two or more different forms); don't be afraid of pauses at the correct moment.

(12) At the same time, try to maintain a minimal amount of fluency, avoiding too many 'ums' or 'ers', and silences resulting from losing your place.

(13) Encourage student notetaking: it can help them focus attention, integrate the information, and will aid later recall.

(14) Consider having a handout available to guide students through the lecture and for reference to salient points later: this could be a list of the main ideas in the lecture or, even better, a series of questions that relate to the material being covered.

(15) Avoid the temptation of including too much content in your presentation: there is no point in 'covering the material' if it is not comprehended.

(16) Remember that attention (and comprehension) will flag after time (some authorities say after only 20–30 minutes); consider giving breaks if your lecture lasts much longer than this.

(17) Try to bear in mind the needs and interests of the audience by relating your ideas to examples from their own experience; if

you are ignorant of their interests and needs, do a little preparatory research or consider asking some probing questions at the start of your presentation.

(18) Remember the previous knowledge and present aptitude of your listeners: again, this may be ascertained by some initial probing questions or even a written quiz if you are working in an educational setting.

(19) Choose an organizational plan for your presentation: for example, a problem-based approach, or a narrative (story-telling) approach.

(20) If an important aim is to stimulate thinking, then consider an open-ended presentation in which you leave certain issues unresolved for the audience to think about later: organization of a lecture that is too complete and precise may not stimulate learning.

(21) The credibility of the presenter is an important factor in influencing opinion: one way of establishing credibility is to demonstrate thorough knowledge of the subject matter, preferably without overtly boasting about it.

(22) At the same time, make sure that your claims of expertise are honest, and be prepared to admit ignorance where appropriate and necessary.

(23) If you are interested in persuading your audience, place the most important information at the beginning; if the audience is intelligent and knowledgeable, include in your presentation counter-arguments to the main thesis, along with an explanation of the limitations of such counter-arguments.

(24) Try to display an enthusiasm for the audience and subject matter: one way of doing this is to make an attempt to discover the particular interests of your listeners.

(25) Consider the introduction of humour, if appropriate, including humour at your own expense; but remember too much hilarity may pre-empt learning.

(26) Try to monitor your audience by watching for signs of interest and attention (eye contact, notetaking, shuffling, etc.).

(27) Eye contact is particularly important in maintaining rapport with your listeners: try to avoid staring at just one or two individuals and instead attempt to sweep the whole of the audience with an occasional glance.

(28) If appropriate, try to obtain feedback by verbal means, such as

asking for comments and questions, or a show of hands to questions that you pose.

(29) Beware of your own physical appearance and its effect on the audience, from the clothes you wear to gestures and mannerisms: have a colleague give comments on this or try recording (on audio- or videotape) a segment of your presentation for the room where you are speaking.

(30) Consider the physical environment for your presentation. Does the lecturn or rostrum form an unnecessary barrier between you and the audience? Could the chairs be arranged differently for more (or less) formality? Can your voice be heard without the use of a microphone? Can those at the back of the hall see any audio-visual aids you may have?

(31) If the audience is small and the auditorium large, consider asking them to rearrange the seating in a more congenial manner.

(32) Be as flexible as possible in your presentation to take account of audience needs (expressed through questions or comments) and any special circumstances that may arise.

(33) You may try to improve your presentations by studying the performance of effective lecturers; but the best way of learning to lecture more effectively is practice and rehearsal, with some type of feedback (via a colleague, videotape or even a mirror) on your performance.

(34) If you suffer badly from nervousness, be especially careful to prepare thoroughly for your presentation; rehearse what you have to say in fairly non-threatening situations (with family or friends) and gradually work up to performing in front of larger audiences; consider alternative presentation strategies that will take you out of the spotlight (audio-visual aids, discussion techniques); and remember that a certain amount of stage fright will boost your arousal level and may improve your performance!

References

Argyle, M. (1978). *The Psychology of Interpersonal Behaviour*. Harmondsworth: Penguin.

_____ (1975). *Bodily Communication*. London: Methuen.

Argyle, M. and Cook, M. (1976). *Gaze and Mutual Gaze*. Cambridge University Press.

Bligh, D. (1972). *What's the Use of Lectures?* Harmondsworth: Penguin.

Bloom, B. S. (1953). Thought processes in lectures and discussions. *J. Gen. Educ.* 7, 160–9.

Brown, J. A. C. (1963). *Techniques of Persuasion: From Propaganda to Brainwashing*. Harmondsworth: Penguin.

Dittman, A. T. (1972). *Interpersonal Messages of Emotion*. New York: Springer.

Dubin, R. and Taveggia, T. C. (1968). *The Teaching–Learning Paradox*. Center for the Advanced Study of Educational Administration, University of Oregon. Monograph No. 18.

Eble, K. (1976). *The Craft of Teaching*. San Francisco: Jossey-Bass.

Feather, N. T. (1969). Preference for information in relation to consistency, novelty, intolerance of ambiguity and dogmatism. *Austr. J. Psychol.* 21, 235–50.

Festinger, L. (1957). *A Theory of Cognitive Dissonance*. New York: Harper and Row.

Giles, H. and Powesland, P. F. (1975). *Speech Style and Social Evaluation*. London: Academic Press.

Goffman, E. (1959). *The Presentation of Self in Everyday Life*. Garden City, NY: Doubleday Anchor.

Greene, E. B. (1928). The relative effectiveness of lecture and individual reading as methods of teaching. *Gen. Psychol. Monogr.* 4, 463–563.

Hartley, J. and Cameron, A. (1967). Some observations on the efficiency of learning. *Educ. Rev.* 20, 30–7.

Hovland, C. I., Janis, I. L. and Kelley, H. H. (1953). *Communication and Persuasion*. Yale University Press.

Howe, M. J. A. and Godfrey, J. (1977). *Student Note Taking as an Aid to Learning*. Exeter University Teaching Services.

Hunt, D. E. and Sullivan, E. V. (1974). *Between Psychology and Education*. Hinsdale, Ill.: Dryden.

Hunter, I. M. L. (1964). *Memory*. Harmondsworth: Penguin.

Katona, G. (1940). *Organizing and Memorizing*. New York: Columbia University Press.

Kelley, H. H. (1950). The warm-cold variable in first impressions of persons. *J. Personality* 18, 431–9.

Knapper, C. K. (1970). The relationship between personality and style of dress. *Bull. Br. Psychol. Soc.* 23, 155–6.

Lewin, K. (1943). Forces behind food habits and methods of change. *Bull. Nat. Res. Council* 108, 36–65.

MacManaway, L. A. (1970). Teaching methods in higher education – innovation and research. *Universities Q.* 24, 321–9.

McKeachie, W. J. (1978). *Teaching Tips: A Guide Book for the Beginning College Teacher*. Lexington, Mass.: D. C. Heath, 7th edn.

McLeish, J. (1976). The lecture method. *In* Gage, N. L. (ed.). *The Psychology of Teaching Methods*. Chicago: National Society for the Study of Education.

Miller, G. A. (1951). *Language and Communication*. New York: McGraw-Hill.

Murray, H. G. and Lawrence, C. (1979). Effect of speech and drama lessons on faculty members' classroom teaching performance. *Reflections*. University of Western Ontario, No. 1, 10–14.

Naftulin, D. H., Ware, J. E. and Donnelly, F. A. (1973). The Doctor Fox lecture: A paradigm of educational seduction. *J. Med. Educ.* 48, 630–5.

Perry, W. G. (1970). *Forms of Intellectual and Ethical Development in the College Years: A Scheme*. New York: Holt, Rinehart and Winston.

Petrie, C. R. (1963). Informative speaking: A summary and bibliography of related research. *Speech Monogr.* 30, 79–91.

Ryan, M. S. (1966). *Clothing: A Study in Human Behavior*. New York: Holt, Rinehart and Winston.

Sargant, W. (1957). *Battle for the Mind*. London: Heinemann.

Scherer, K. R. (1971). Attribution of personality from voice: A cross-cultural study on interpersonal perception. *Proceedings of the 79th Annual Convention of the American Psychological Association*, pp. 351–2.

Secord, P. F. (1958). Facial features and inference processes in interpersonal perception. *In* Tagiuri, R. and Petrullo, L. (eds). *Person Perception and Interpersonal Behavior*. Stanford University Press, pp. 300–15.

Shannon, C. and Weaver, W. (1949). *The Mathematical Theory of Communication*. Urbana, Ill.: University of Illinois Press.

Sheffield, E. F. (ed.) (1974). *Teaching in the Universities: No one way*. Montreal: McGill-Queen's University Press.

Smith, B. L., Lasswell, H. D. and Casey, R. D. (eds) (1946). *Propaganda, Communication and Public Opinion*. Princeton University Press.

Thompson, W. N. (1967). *Quantitative Research in Public Address and Communication*. New York: Random House.

Warr, P. B. and Knapper, C. K. (1968). *The Perception of People and Events*. London: Wiley.

Wiseman, G. and Barker, L. (1967). *Speech: Interpersonal Communication*. San Francisco: Chandler.

7 Inter-cultural communication

MICHAEL ARGYLE

Introduction

Many people have to communicate and work with members of other cultures, and social skills training is now being given to some of those who are about to work abroad. Inter-cultural communication (ICC) is necessary for several kinds of people. Tourists are probably the largest category, though they stay for the shortest periods and need to master skills in only a few simple situations: meals, travel, shopping, taxis, etc. They are largely shielded from the local culture by the international hotel culture. Business, governmental and university visitors, on short business trips, have to cope with a wider range of problems but are often accommodated in hotels or somewhere similar, and looked after by other expatriates. They, too, are somewhat shielded from the local culture; they rarely learn the language, and are given a great deal of help. Businessmen or others on longer visits of up to 5 years, students who stay 1−3 years, and members of the Peace Corps and Voluntary Service Overseas who stay for 2 years are in a much more demanding situation: living in a house or apartment, coping with many aspects of the local culture and learning at least some of the language. Immigration may take place as a deliberate move, or as a gradual process while a visit becomes extended. This requires mastery of the new culture, as well as changes of attitude and self-image. Those who stay at home may meet visitors from abroad and may need to work effectively with them. They may also have to deal with refugees, those from other racial groups and other social classes. However these contacts are usually limited to meals and work settings.

A number of category schemes have been produced to describe the main modes of response of visitors to different cultures. The principal alternatives are: detached observers, who avoid participation; reluctant and cautious participants in the local culture; enthusiastic participants, some of whom come to reject their original culture; and settlers (Brein and David 1971).

How can intercultural effectiveness be assessed? An important minimal criterion is whether an individual manages to complete the planned tour or whether he packs up and returns home early. Over 60 per cent of Peace Corps members in some areas, and a similar proportion of British businessmen from some firms posted to Africa, fail to complete their tours, often at great cost to their firms. For those who succeed in staying the course there are several possible indices of success:

(1) Subjective ratings of comfort and satisfaction with life in the other culture (e.g. Gudykunst, *et al.* 1977).

(2) Ratings by members of the host culture of the acceptability or competence of the visitor (e.g. Collett 1971).

(3) Ratings by the field supervisor of an individual's effectiveness at the job, as has been used in Peace Corps studies. The effectiveness of salesmen could be measured objectively, and this applies to a number of other occupational roles.

(4) Performance in role-played inter-cultural group tasks, as used by Chemers, *et al.* (1966).

Hammer, *et al.* (1978) analysed ratings by returned visitors to other cultures and found that they recognized three dimensions of inter-cultural competence: ability to deal with psychological stress; ability to communicate effectively; and ability to establish interpersonal relations.

Competent performance as a visitor to another culture, or in dealing with members of another culture, can be regarded as a social skill, analogous to the skills of teaching, interviewing, etc. ICC is different in that a very wide range of situations and types of performance are included, together with a variety of goals. Inter-cultural skills may include some quite new skills, where quite different situations or rules are encountered, such as bargaining or special formal occasions. It may be necessary to perform familiar skills in a modified style, e.g. a more authoritarian kind of supervision, or more intimate social relationships. There are often a number of themes or modes of interaction in a culture, which are common to a wide range of situations. I suggest that these themes can be the most useful focus of training for ICC. The next section examines the main themes of this kind.

There is a special phenomenon here which has no clear equivalent among other social skills, i.e. 'culture shock'. Oberg (1960) used this term to refer to the state of acute anxiety produced by unfamiliar social norms and social signals. Others have extended the notion to include the fatigue of constant adaptation, the sense of loss of familiar food, companions, etc., rejection of the host population or rejection by it, confusion of values or identity, discomfort at violation of values and a feeling of incompetence at dealing with the environment (Taft 1977).

Some culture shock is common among those living abroad for the first time, especially in a very different culture, and it may last 6 months or longer. Those going abroad for a limited period, like a year, show a U-shaped pattern of discomfort: in the first stage they are elated, enjoy the sights and are well looked after. In the second stage they have to cope with domestic life, and things get more difficult; they keep to the company of expatriates and are in some degree of culture shock. In the third phase they have learnt to cope better and are looking forward to returning home. There may be problems when they do return home, and many people experience problems of re-entry owing to, for example, a loss of status or a less exciting life (Brein and David 1971).

Furnham and Bochner (in press) studied 400 foreign students from fifty-seven countries in England. Greater difficulty was reported by students from more distant lands – the East followed by Southern Europe and Northern Europe. Factor analysis of scores for difficulty in forty situations produced six factors; the factors most affected by cultural distance were: formal situations and being the focus of attention; managing intimate relationships; initiating social contacts and being introduced to new people. The most difficult individual situations for foreign students in Britain were concerned with establishing and maintaining personal relationships with the British.

Another special problem for ICC is how far a visitor should accommodate to local styles of behaviour. Europeans and Americans in Africa and Third World countries usually find that they are *not* expected to wear local clothes or engage in exotic greetings. There seems to be a definite 'role of the visitor' to which one is expected to conform. Rather greater accommodation to local ways, which may include mastering the language, is expected of those who stay for longer periods. In the USA, on the other hand, much greater conformity is expected, probably as a result of the long history of assimilating immigrants. Where total conformity is not required, visitors are still expected to show a positive attitude towards the local culture and not complain or criticize, like the so-called 'Whinging Pom' in Australia. There may be a temptation to keep to hotels, clubs and cantonment, but this will lead to isolation from the local community. Bochner, *et al.* (1977) found that foreign students usually had friends both from their home country and from the country they were in − the latter were needed to help them cope with the culture.

In this chapter I shall examine some of the areas of difference between cultures which can give rise to communication problems. Any successful form of social skills training for ICC should take account of these differences. Then I shall discuss the main forms of training which have been developed for this purpose.

Cultural differences in social interaction

Language

This is one of the most important differences between many cultures, and one of the greatest barriers.

The person who has learnt a language quite well can still make serious mistakes, as with the Dutchman on a ship who was asked if he was a good sailor and replied indignantly that he was not a sailor but a manager.

Several studies have shown that language fluency is a necessary condition for the adjustment of foreign students in the USA, though there is also evidence that confidence in the use of language regardless of ability is just as important (Gullahorn and Gullahorn 1966). Often there are variations in accent, dialect and grammar − as in Black-American English − or in the language used − as in multilingual communities. An individual may indicate a positive or negative attitude to another by shifting towards a more similar or less similar speech style (Giles and Powesland 1975). Visitors to another culture should be aware of the impression they are creating by the speech style which they use. While efforts to speak the language are usually well received, this is not always so; the French dislike the inaccurate use of their language. Taylor and Simard (1975) found that lack of inter-action between English and French Canadians was caused less by lack of language skills than by attitudes; language helped to preserve ethnic identity.

Most cultures have a number of forms of polite usage, which may be misleading. These may take the form of exaggeration or modesty. Americans ask questions which are really orders or requests ('Would you like to . . .?'). And in every culture, in many situations, there are special forms of words, or types of conversation, which are thought to be appropriate — e.g. to ask a girl for a date, to disagree with someone at a committee, to introduce people to each other and so on. Americans prefer directness, but Mexicans regard openness as a form of weakness or treachery, and think one should not allow the outside world to know their thoughts. Frankness by Peace Corps volunteers in the Philippines leads to disruption of smooth social relationships (Brein and David 1971).

There are cultural differences in the sequential structure of conver-sations. The nearly universal question–answer sequence is not found in some African cultures where information is precious and not readily given away (Goody 1978). In Asian countries the word 'no' is rarely used, so that 'yes' can mean 'no' or 'perhaps'. Saying 'no' would lead to loss of face by the other, so indirect methods of conveying the message may be used, such as serving a banana (an unsuitable object) with tea to indicate that a marriage was unacceptable (Cleveland 1960). The episode structure of conversations varies a lot: Arabs and others have a 'run-in' period of informal chat for about half-an-hour before getting down to business.

Some of these differences are due to different use of non-verbal (NV) signals. Erickson (1976) found that white Americans interview-ing blacks often thought the interviewee wasn't attending or under-standing because they did not look while listening, and kept rewording questions in simpler and simpler forms. In several cultures 'thank-you' is signalled non-verbally; in China this is done at meals by rapping lightly on the table.

Non-verbal communication (NVC)

NVC plays several essential roles in social interaction: communicating attitudes to others, e.g. of like–dislike, expressing emotions, and supporting speech by elaborating on utterances, providing feedback from listeners, and governing synchronization. Although NV signals are used in similar ways in all cultures, there are also differences and these can easily produce misunderstanding (Argyle 1975). Triandis, *et al.* (1968) observed that friendly criticism may be interpreted as hatred, and very positive attitudes as neutral, by someone from another culture. Several studies have found that if people from culture A are trained to use the NV signals of culture B (gaze, distance, etc.), they will be liked more by members of the second culture (e.g. Collett 1971).

The face is the most important source of NVC. Similar basic emotional expressions are found in all cultures, and are at least partially innate. However, Chan (1979) has found that the Chinese express anger and disgust by narrowing the eyes, the reverse of that found in the USA. There are also different display rules, prescribing when these expressions may be shown, when one may laugh, cry and so on (Ekman, *et al.* 1972). We carried out an experiment on the inter-cultural communication of interpersonal attitudes, in which judges decoded videotapes, the main cues being face and voice. As Table 7.1 shows, Japanese subjects found it easier to decode English and Italian than Japanese performers, probably because Japanese display rules forbid the use of negative facial expressions (Shimoda, *et al.* 1978).

Table 7.1 Accuracy of recognition of NV cues for emotions and interpersonal attitudes by English, Italian and Japanese (Results in percentages).*

	Performers			
	English	*Italian*	*Japanese*	*Average*
Judges:				
English	60.5	55	36	50
Italian	52	61.5	29	47
Japanese	54	56	43	51
Average	56	57	36	

* From Shimoda, *et al.* 1978.

This shows that the Japanese are indeed relatively 'inscrutable', but it is not yet known whether they make use of alternative channels such as posture for transmitting information normally conveyed by the face. There are also some variations of facial expression within cultures, between different regions and social classes. Seaford (1975) reports the use of a 'pursed smile' facial dialect in the State of Virginia.

Gaze is also similarly used in all cultures but the amount of gaze varies quite widely. Watson (1970) studied the gaze of pairs of students from different countries, with the results shown in Table 7.2. The highest levels of gaze (lowest scores in table) were shown by Arabs and Latin Americans, the lowest by Indians and Northern Europeans (highest scores). When people from different cultures met, if the other had a low level of gaze he was seen as not paying attention, impolite or dishonest; while too much gaze was seen as disrespectful, threatening or insulting. Some cultures have special rules about gaze, such as not looking at certain parts of the body or at certain people. Gaze may have a special meaning, as when old ladies with squints are believed to have the Evil Eye (Argyle and Cook 1976).

Table 7.2 Scoring of levels of gaze between pairs of students from different countries.*

Culture type	Groups	No.	Mean score	SD
Contact	Arabs	29	2.57	±1.15
	Latin Americans	20	2.47	±1.01
	Southern Europeans	10	2.19	±0.59
Non-contact	Asians	12	3.25	±0.49
	Indians–Pakistanis	12	3.59	±0.68
	Northern Europeans	48	3.51	±0.99

SD = standard deviation from mean.
Scoring system: 1 = sharp (focusing directly on other person' eyes); 2 = clear (focusing about other person's head and face); 3 = peripheral (having other person within field of vision but not focusing on head or face); 4 = no visual contact (looking down or gazing into space).
* From Watson (1970).

Spatial behaviour varies between cultures. Watson and Graves (1966) confirmed earlier observations that Arabs stand much closer than Americans (or Western Europeans), and found that they also adopt a more directly facing orientation. When an Arab and an American meet it would be expected that the American would move backwards, turning in a backwards spiral, closely followed by the Arab. An elaborate set of rules about distance is found in India, prescribing exactly how close members of each caste may approach other castes. There are also rules for spatial behaviour in different situations – far greater crowding is allowed in lifts and buses, football matches and sherry parties. There are other cultural differences in the use of space. Americans establish temporary territorial rights in public places, but Arabs do not consider that people have such rights – e.g. to the seat they are sitting on.

or example for consuming alcohol in some Arab countries. There are ules about how the eating is performed — with knife and fork, hopstick, right hand, etc. There are also extensive rules about table manners — when to start eating, how much to leave, how to obtain or efuse a second helping and so on.

Rules about time   How late is 'late'? This varies greatly. In Britain nd North America one may be 5 minutes late for a business appointment, but not 15 and certainly not 30 minutes late, which is perfectly normal in Arab countries. On the other hand in Britain it is correct to e 5–15 minutes late for an invitation to dinner. An Italian might rrive 2 hours late, an Ethiopian later and a Javanese not at all — he ad accepted only to prevent his host losing face (Cleveland, *et al.* 960). A meal in Russia at a restaurant normally takes at least 3 ours. In Nigeria it may take several days to wait one's turn at a overnment office, so professional 'waiters' do it for you.

eating guests   In Britain, in middle-class circles at least, there are ules about seating people at table, when there are six, eight or other umbers present. In the USA there appear to be no such rules, and ritish visitors are commonly surprised to see familiar rules broken. n China the tables are circular and the seating rules are different gain, and similar to the British though the most important person aces the door. In Japan different seating positions in a room have ifferent status. There may also be rules about who should talk to hom, as in the 'Boston switch': hostess talks to person on her right uring first course, switches to person on her left for the next course nd everyone else pairs off accordingly.

Rules based on ideas   Sometimes the rules of another culture are uite incomprehensible until one understands the ideas behind them. n Moslem countries there are strict rules based on religious ideas, uch as fasting during Ramadan, saying prayers five times each day, nd giving one fortieth of income as alms (Roberts 1979). To visit ome kinds of Australian aboriginals it is necessary to sit at the edge of heir land and wait to be invited further: to move closer would be egarded as invasion of territory. It is necessary for them to have noking fires (without chimneys) for religious reasons, despite ossible danger to the health of those inside (O'Brien and Plooij 977).

In addition to different rules for the same or similar situations there may also be new situations. Black-American youths play the 'dozens' (ritual insulting of the other's mother), other Americans go on picnics, Chinese families go to pay respect to their ancestors, Oxford dons drink port and take a special form of dessert. There may be special ceremonies connected with engagement, marriage, childbirth and other rites of passage.

Cultures also vary in the extent to which behaviour is a function of situations, as a result of their rules and other properties. Argyle, *et al.* (1978) found that Japanese were more influenced by situations, while the British behaved more consistently, i.e. as a function of personality. This means that it is more difficult to infer the properties of personality from instances of behaviour in the Japanese than in the British.

Within cultures in developing countries there are often two sets of rules and ideas, corresponding to traditional and modern attitudes. Inkeles (1969) found similar patterns of modernization in different countries, centred around independence from parental authority, concern with time, inclusion in civil affairs and openness to new experience. Dawson, *et al.* (1971) devised T—M scales, of which some of the core concepts were attitudes to parental authority, gift-giving and the role of women. Modernism is highly correlated with education and social class.

In some cases it is essential for the visitor to conform to rules, for example in matters of eating and drinking. In other cases the rules may be in conflict with his own values, the practice of his home organization, or the laws of his own country, as in the case of 'bribery'. There may be no straightforward solution to these problems, but it is at least necessary to recognize what the local rules are, and the ideas behind them, rather than simply condemning them as wrong.

Social relationships

The pattern of social relationships at work, in the family and with friends takes a somewhat different form in different cultures, and different skills are needed to handle these relationships. Surveys by Triandis, *et al.* (1968) and other research workers have shown that relationships vary along the same dimensions in all cultures: in group/out-group, status, intimacy and hostility or competition.

Family relationships In developing countries the family is more important than in developed countries. A wider range of relatives are actively contacted; relationships are closer and greater demands are made. These include helping to pay for education, helping to get jobs, and helping when in trouble. Foa and Chemers (1967) point out that in traditional societies the family is the most important source of relationships and many different role relationships are distinguished, but there are relatively few outside the family. Throughout Africa and the Middle East the family takes a similar form: marriage is arranged as a contract, and money (or sometimes camels) is paid for the bride, kinship is traced through the father and male relatives and polygyny is accepted (Roberts 1979). In China great respect is paid to older generations: parents are respected, large financial contributions are made to the family by unmarried children who have left home, regular visits are paid to the graves of ancestors. The family itself may take varied forms, such as having more than one wife, or a wife and concubines.

The way in which different relations are grouped as similar varies: distinctions may be based primarily on age, generation, consanguinity or sex (Tzeng and Landis 1979). Sex roles vary: in the Arab world women traditionally do not work or drive cars but spend most of their time at home. The reverse operates in countries like Israel, China and Poland where women do nearly all the same jobs as men. Patterns of sexual behaviour vary: promiscuity may be normal, or virginity greatly prized; businessmen visiting parts of the East are sometimes embarrassed by being offered girls as part of the hospitality. Cultures vary from complete promiscuity before and after marriage to a complete taboo on sex outside marriage (Murdoch 1949). Goody (1976) has shown that there is great control over premarital sexual behaviour in societies which have advanced agriculture, where marriage is linked with property (especially land) transactions so that unsuitable sexual attachments must be controlled. Americans, and to a lesser extent Europeans, mix work and family life and receive business visitors into the home; Japanese and Arabs do not.

Supervision of groups In most of the world outside Europe and North America there is greater social distance between ranks, more deference and obedience and a generally more authoritarian social structure. Subordinates do not speak freely in front of more senior people, and less use is made of face-to-face discussion. Melikian (1959) found that Egyptian Arabs, whether Moslem or Christian, had higher scores on authoritarianism than Americans. While the democratic-persuasive style of leadership is most effective in the USA and Europe, this is not the case elsewhere. In India the authoritarian style has been found to be more effective; in China there was no difference and in Japan authoritarian-led groups did best with a difficult task (Mann 1980). In Japan teachers and superiors at work adopt an *Oyabun-Koyun* relationship, consisting of a paternalistic care for subordinates.

Groups   Ethnographic studies have shown that groups have more power over their members in a number of cultures than in others — in Japan, China, Israel and Russia, for example. The individual is subordinated more to the group, and a high degree of conformity is expected. Americans and Europeans are thought to be more individualistic, and social psychological experiments have shown relatively low levels of conformity in Germany and France. Conformity pressures are stronger in the cultures where conformity is greatest. In Japan group decisions are traditionally carried out by a kind of acquiescence to the will of the group, without voting. In some cultures there is great stress on co-operation rather than competition in groups, e.g. in the Israeli kibbutz, Mexican villages and among Australian aboriginals (Mann 1980).

Castes and classes   In all cultures there are hierarchical divisions of status and horizontal divisions of inclusion and exclusion. The hierarchical divisions may take the form of social classes, which can be recognized by clothes and accent (as in Britain) or in other ways. There may be ethnic groups which have their places in the hierarchy, as in the USA. Or there may be immutable castes, as in India. This creates special problems for visitors to India: Western visitors are relatively rich and clean, and so appear to be of high caste, but also eat meat even with the left hand and drink alcohol like untouchables, so a special visitor caste of *videshis* has been created. However, visitors to ashrams who adopt the costume of holy men do not fit this caste and cause great offence to the Indians (Wujastyk 1980). The horizontal divisions between different tribes or class are also of great importance. In Africa it may be necessary to make up work groups from members of the same tribe, and it would be disastrous to appoint a leader from another tribe. Similar clan divisions are of course found in Scotland, and also in China (Hsu 1963). In-group versus out-group distinctions can take varied forms. Studies of helping behaviour have found that fellow countrymen are usually given more help than visitors, but in Greece tourists are treated like family and friends (Triandis, *et al.* 1968).

Motivation

Several forms of motivation have been found to differ on average between cultures. This means that typical members of another culture are pursuing different goals and are gratified by different rewards. Sometimes the causes of these motivational differences can be found in other features of a culture. For example, members of societies which are constantly at war with their neighbours encourage aggressiveness in their young males (Zigler and Child 1969). In some cultures there is almost no interest in wealth or material possessions.

Achievement motivation   McClelland (1961) found that cultures differed in the level of achievement motivation, as measured by the popularity of children's stories with achievement themes; the high need for achievement (NAch) countries had higher rates of economic growth, and this may be partly due to the motivational difference. The USA over the last century has had a very high NAch; underdeveloped countries have been lower. McClelland and Winter (1969) ran a training course for Indian managers, in which the latter role-played high NAch managers. The result was that they increased the size and turnover of their enterprises after attending the course. There is of course a wide range of individual differences within a culture, but it is worth realizing that in some areas individuals are likely to work hard and take risks in order to earn more money, improve their status and to build up the enterprise in which they work; whereas in other areas people expect to be rewarded on the basis of the social position of their family or clan, not on their own efforts.

Assertiveness   Assertiveness, or dominance as opposed to submissiveness, is one of the main dimensions along which social behaviour varies. In the USA social skills training has concentrated on assertiveness, presumably reflecting a widespread approval of and desire to acquire assertive behaviour. This interest in assertiveness is strong among American women as part of the women's movement. It has also been suggested that the absence of universally accepted rules makes it necessary to stand up for your rights rather often.

Americans are perceived as assertive in other parts of the world. However, there are some cultures, e.g. China and part of Indonesia, where assertiveness is not valued highly, and submissiveness and the maintenance of pleasant social relations are valued more (Noesjirwan 1978). In Britain the candidates for social skills training are more interested in making friends. Furnham (1979) found that European white nurses in South Africa were the most assertive, followed by Africans and Indians.

Extraversion   Surveys using extraversion questionnaires show that Americans and Canadians are more extraverted than the British (e.g. Eysenck and Eysenck 1969). What exactly this means in terms of social behaviour is rather unclear. It is commonly observed that Americans are good at the early stages of a relationship, where the British can be shy and awkward. In the USA the peer group plays an important part in the life of children and adolescents; among adults great value is placed on informal relationships (Riesman, *et al.* 1955).

In the East great value is placed on maintaining good social relationships so that assertiveness and disagreement are avoided. However, this can be confined to members of the same family, clan or group.

Face   In Japan, and to a lesser extent other parts of the Far East, maintaining face is highly important. Special skills are required to make sure that others do not lose face. Foa, *et al.* (1969) found that students from the Far East who experienced failure in an experimental task withdrew from whoever told them of their failure. In negotiations it may be necessary to make token concessions before the other side can give way. Great care must be taken at meetings over disagreeing or criticizing, and competitive situations should be avoided.

Concepts and ideology

Certain aspects of life in another culture may be incomprehensible without an understanding of the underlying ideas. Some of these ideas are carried by language, and knowing a language deepens understanding of the culture. The words in a language reflect and provide labels for the cognitive categories used in the culture to divide up the world. The colour spectrum is divided up in different ways, and the colour words reflect this, in different cultures (Berlin and Kay 1969).

The same is true of every other aspect of the physical and social world, so that knowledge of the language provides knowledge of the culture. Translation of words may lead to changes of emotional association: the Australian word 'Pom' does not only mean 'British immigrant' but has negative and joking associations as well. Words in one language and culture may have complex meanings which are difficult to translate, as with the Israeli Chutzpah ('outrageous cheek', such as exporting tulips to Holland), Russian and western concepts of 'freedom' and 'democracy', and the Japanese concept of the *Oyabum—koyum* relationship (see p. 184).

There may be misunderstandings due to differences of thinking. Sharma (unpublished observations) notes how western observers have criticized Indian peasants for their passivity and general lack of the Protestant ethic, whereas they have produced a great increase in productivity by adapting to the Green Revolution. African languages are often short of words for geometrical shapes, so that it is difficult to communicate about spatial problems. Some words or ideas may be taboo, e.g. discussion of family planning (Awa 1979).

Some of the differences in rules which were discussed above can be explained in terms of the ideas behind them, as in 'bribery' and 'nepotism'. Attitudes to business practices are greatly affected by ideas and ideology. Marxists will not discuss 'profits'; Moslems used to regard 'usury' as sinful. Surprisingly, the stricter forms of Protestantism have been most compatible with capitalism and gave rise to the 'Protestant ethic' (Argyle 1972).

Values These are broader, more abstract goals; the general states of affairs which are regarded as desirable. Triandis (1972) studied twenty values by asking for the antecedents and consequences of eleven concepts. In parts of India they found that status and glory were valued most, whilst wealth was not valued (being associated with arrogance and fear of thieves), nor was courage or power. The Greeks valued punishment (which was associated with justice) and power. The Japanese valued serenity and aesthetic satisfaction, and disvalued ignorance, deviation and loneliness. Szalay and Maday (cited by Triandis, *et al.* 1972) found that Americans rated love and friendship as their most important life concerns, health as fifth: Koreans ranked these values as twelfth, fourteenth and nineteenth. Triandis (1971) found that 'work' was regarded as a good thing in moderately difficult environments where economic development was rapid, but it was rated less favourably in easy or difficult environments.

Training methods

Language learning

There are many cultures where visitors, especially short-term visitors, can get by quite well without learning the language. On the other hand, this probably means that they are cut off from communicating with most of the native population and that they do not come to understand fully those features of the culture which are conveyed by language. Language learning can be greatly assisted by the use of a language laboratory and textbooks such as that by Leech and Svartlik (1975), where such books are available, which provide detailed information on the everyday informal use of language.

Use of educational methods

Despite the preference for more active methods of SST in other areas, for ICC reading and lectures are currently the most widely used methods. The most sophisticated approach here has been the development of Culture Assimilators. Critical-incident surveys of occasions when Americans have got into difficulty in Thailand, Greece, etc. have been conducted and a standard set of difficult episodes has been written, for example:

'One day a Thai administrator of middle academic rank kept two of his assistants about an hour from an appointment. The assistants, although very angry, did not show it while they waited. When the administrator walked in at last, he acted as if he were not late. He made no apology or explanation. After he was settled in his office, he called his assistants in and they all began working on the business for which the administrator had set the meeting.'

(Brislin and Pedersen 1976, pp. 90–1.)

Several explanations are offered, of which the correct one is: 'In Thailand, subordinates are required to be polite to their superiors, no matter what happens, nor what their rank may be' (ibid., p. 92), and further information is added. These episodes are put together in a tutor text, which students work through by themselves (Fiedler, *et al.* 1971).

There have been several follow-up studies of the use of culture assimilators, showing modest improvements in handling mixed cultural groups in laboratory settings, and in one case in a field setting. However, not very much field assessment has been done, the effects of training have not been very striking and the subjects used have all been of high motivation and intelligence (Brislin and Pedersen 1976).

A similar method is the use of case studies. These are widely used for management training in international firms, the cases being based on typical managerial problems in the other culture. They play an important part in 2-week courses, using educational methods. It is common to include wives and children in such courses, with special materials for them too (DiStephano 1979).

Educational methods can probably make a valuable contribution to cross-cultural training, since there is always a lot to learn about another culture. However, as with other skills such intellectual learning must be combined with practice of the skills taught.

Role playing

Several forms of role playing have been used for ICC, though it has not been the usual form of training. One approach is to train people in laboratory situations in the skills or modes of communication of a second culture, using videotape playback. Collett (1971) trained Englishmen in the NV communication styles of Arabs, and found that those trained in this way were liked better by Arabs than were members of a control group.

The American Peace Corps has used simulation techniques to train their members. Trainees have been sent to work on an American-Indian reservation for example. Area simulation sites were constructed to train members for different locations, e.g. one in Hawaii for South-east Asia volunteers, complete with water buffaloes. However, it is reported that these rather expensive procedures have not been very successful (Brislin and Pederson 1976), and they have been replaced by training in the second culture itself.

Interaction with members of the other culture

In the Intercultural Communication workshop trainees go through several exercises with members of the other culture, and use is made of role playing and the study of critical incidents (Alther 1975). This looks like a very powerful method, but no follow-up results are available. At Farnham Castle in Britain the training courses include meetings with members of the other culture, and with recently returned expatriates.

When people arrive in a new culture they are often helped both by native members of the culture and by expatriates. Bochner, *et al.* (1977) found that foreign students in Hawaii usually had friends from both of these groups, who could help them in different ways.

Combined approaches

Each of the methods described has some merits, and a combination of methods would probably be the most effective. This might include some language instruction, learning about the other culture, role playing and interaction with native members of the culture. Gudykunst, *et al.* (1977) used a combination of several methods, though not including any language teaching, in a 3-day course, and found that this led to higher reported levels of satisfaction for Naval personnel posted to Japan.

Guthrie (1966) describes one of the training schemes used by the Peace Corps for those going to the Philippines. The training included:

(1) Basic linguistics, so that trainees could pick up local dialects quickly; later this was replaced by teaching specific dialects.
(2) Lectures by experts on different aspects of the Philippine culture.
(3) Physical and survival training at the Puerto-Rican jungle camp; as noted earlier this was replaced by training in the culture itself.

Conclusions

Many people go abroad to work in other cultures; some of them fail to complete their mission and others are ineffective, because of difficulties with ICC.

Difficulties of social interaction and communication arise in several main areas:

(1) Language, including forms of polite usage.
(2) NV communication: uses of facial expression, gesture, proximity, touch, etc.

(3) Rules of social situations, e.g. for bribery, gifts and eating.
(4) Social relationships: within the family, at work and between members of different groups.
(5) Motivation, e.g. achievement motivation and for face saving.
(6) Concepts and ideology, e.g. ideas derived from religion and politics.

Several kinds of training for ICC have been found to be successful, especially in combination. These include language learning, educational methods, role playing and interaction with members of the other culture.

References

Alther, G. L. (1975). Human relations training and foreign students. *In* Hoopes, D. (ed.). *Readings in Intercultural Communication*. Pittsburgh: Intercultural Communications Network of the Regional Council for International Education, vol. 1.

Argyle, M. (1972). *The Social Psychology of Work*. Harmondsworth: Penguin Books.

———— (1975). *Bodily Communication*. London: Methuen.

Argyle, M. (ed.) (1981). *Social Skills and Health*. London: Methuen.

Argyle, M. and Cook, M. (1976). *Gaze and Mutual Gaze*. Cambridge University Press.

Argyle, M., Furnham, A. and Graham, J. A. (eds) (1981). *Social Situations*. Cambridge: Cambridge University Press.

Argyle, M., Shimoda, K. and Little, B. (1978). Variance due to persons and situations in England and Japan. *Bri. J. Soc. Clin. Psychol.* 17, 335–7.

Awa, N. E. (1979). Ethnocentric bias in developmental research. *In* Asante, M. K., Newmark, E. and Blake, C. A. (eds). *Handbook of Intercultural Communication*. Beverly Hills and London: Sage, pp. 263–81.

Berlin, B. and Kay, P. (1969). *Basic Color Terms*. Berkeley, Cal.: University of California Press.

Bochner, S., McLeod, B. M. and Lin, A. (1977). Friendship patterns of overseas students: a functional model. *Int. J. Psychol.* 12, 277–94.

Brein, M. and David, K. H. (1971). Intercultural communication and the adjustment of the sojourner. *Psychol. Bull.* 76, 215–30.

Brislin, R. W. and Pedersen, P. (1976). *Cross-Cultural Orientation Programs*. New York: Gardner Press.

Chan, J. (1979). *The Facial Expressions of Chinese and Americans*. Ph.D. thesis, South Eastern University, Louisiana.

Chemers, M. H., Fiedler, F. E., Lekhyananda, D. and Stolurow, L. M. (1966). Some effects of cultural training on leadership in heterocultural task groups. *Int. J. Psychol.* 1, 301–14.

Cleveland, H., Mangone, G. J. and Adams, J. G. (1960). *The Overseas Americans*. New York: McGraw-Hill.

Collett, P. (1971). On training Englishmen in the non-verbal behaviour of Arabs: an experiment in intercultural communication. *Int. J. Psychol.* 6, 209–15.

Dawson, J., Whitney, R. E. and Lan, R. T. S. (1971). Scaling Chinese traditional–modern attitudes and the GSR measurement of 'Important' versus 'Unimportant' Chinese concept. *J. Cross-Cult. Psychol.* 2, 1–27.

DiStephano, J. J. (1979). Case methods in international management training. *In* Asante, M. K., Newmark, E. and Blake, C. A. (eds). *Handbook of Intercultural Communication.* Beverly Hills and London: Sage, pp. 421–46.

Ekman, P., Friesen, W. V. and Ellsworth, P. (1972). *Emotion in the Human Face: Guildelines for Research and a Review of Findings.* New York: Pergamon Press.

Erickson, F. (1976). Talking down and giving reasons: hyper-explanation and listening behavior in inter-social situations. Presented at Ontario institute for the Study of Education Conference, Toronto.

Eysenck, H. J. and Eysenck, S. B. G. (1969). *Personality Structure and Measurement* London: Routledge and Kegan Paul.

Fiedler, F. E., Mitchell, R. and Triandis, H. C. (1971). The culture assimilator: an approach to cross-cultural training. *J. Appl. Psychol.* 55, 95–102.

Foa, U. and Chemers, M. (1967). The significance of role behavior differentiation for cross-cultural interaction training. *Int. J. Psychol.* 2, 45–57.

Foa, U. G., Mitchell, T. R. and Lekhyananda, D. (1969). Cultural differences in reaction to failure. *Int. J. Psychol.* 4, 21–6.

Furnham, A. (1979). Assertiveness in three cultures: multidimensionality and cultural differences. *J. Clin. Psychol.* 35, 522–7.

Furnham, A. and Bochner, S. (1981). Social difficulty in a foreign culture: the difficulties reported by foreign students in everyday social situations in England. *In* Argyle, M., Furnham, A. and Graham, J. A. (eds). *Social Situations.* Cambridge: Cambridge University Press, pp. 344–57.

Giles, H. and Powesland, P. F. (1975). *Speech Style and Social Evaluation.* London: Academic Press.

Goody, E. N. (1978). Towards a theory of questions. *In* Goody, E. N. (ed.). *Questions and Politeness.* Cambridge University Press.

Goody, J. (1976). *Production and Reproduction.* Cambridge University Press.

Graham, J. A. and Argyle, M. (1975). A cross-cultural study of the communication of extra-verbal meaning of gestures. *Int. J. Psychol.* 10, 57–67.

Gudykunst, W. B., Hammer, M. R. and Wiseman, R. L. (1977). An analysis of an integrated approach to cross-cultural training. *Int. J. Intercult. Rel.* 1, 99–110.

Gullahorn, J. E. and Gullahorn, J. T. (1966). American students abroad: professional versus personal development. *The Annals* 368, 43–59.

Guthrie, G. M. (1966). Cultural preparation for the Philippines. *In* Textor, R. B. (ed.). *Cultural Frontiers of the Peace Corps.* Cambridge, Mass.: MIT Press.

Hammer, M. R., Gudykunst, W. B. and Wiseman, R. L. (1978). Dimensions of intercultural effectiveness: an exploratory study. *Int. J. Intercult. Relat.* 2, 382–393.

Hewes, G. (1957). The anthropology of posture. *Sci. Am.* 196, 123–32.

Hsu, F. L. K. (1963). *Caste, Clan and Club*. Princeton, NJ: Van Nostrand.

Inkeles, A. (1969). Making men modern: on the causes and consequences of individual change in six developing countries. *Am. J. Sociol.* 75, 208–25.

Krout, M. H. (1942). *Introduction to Social Psychology*. New York: Harper and Row.

Leech, G. and Svartlik, J. (1975). *A Communicative Grammar of English*. London: Longman.

McClelland, D. C. (1961). *The Achieving Society*. Princeton, NJ: Van Nostrand.

McClelland, D. C. and Winter, D. G. (1969). *Motivating Economic Achievement*. New York: Free Press.

Mann, L. (1980). Cross cultural studies of small groups. *In* Triandis, H. (ed.) *Handbook of Cross Cultural Psychology*. Reading, Mass.: Addison-Wesley, vol 5.

Melikian, L. H. (1959). Authoritarianism and its correlation in the Egyptian culture and in the United States. *J. Soc. Issues* 15, 58–68.

Morris, D., Collett, P., Marsh, P. and O'Shaughnessy, M. (1979). *Gestures: their Origins and Distribution*. London: Cape.

Morsbach, H. (1977). The psychological importance of ritualized gift exchange in modern Japan. *Ann. N.Y. Acad. Sci.* 293, 98–113.

Murdoch, G. P. (1949). *Social Structure*. New York: Macmillan.

Noesjirwan, J. (1978). A rule-based analysis of cultural differences in social behaviour: Indonesia and Australia. *Int. J. Psychol.* 13, 305–16.

Oberg, K. (1960). Cultural shock: adjustment to new cultural environments. *Pract. Anthropol.* 7, 177–82.

O'Brien, G. E. and Plooij, D. (1977). Development of culture training manuals for medical workers with Pitjantjatjara Aboriginals. *J. Appl. Psychol.* 62, 499–505.

Riesman, D., Glazer, N. and Denney, R. (1955). *The Lonely Crowd: a Study of the Changing American Character*. New York: Doubleday.

Roberts, G. O. (1979). Terramedian value systems and their significance. *In* Asante, M. K., Newmark, E. and Blake, C. A. (eds). *Handbook of Intercultural Communication*. Beverly Hills and London: Sage, pp. 203–08.

Saitz, R. L. and Cervenka, E. J. (1972). *Handbook of Gestures: Columbia and the United States*. The Hague: Mouton.

Seaford, H. W. (1975). Facial expression dialect: an example. *In* Kendon, A., Harris, R. M. and Key, M. R. (eds). *Organization of Behavior in Face-to-Face Interaction*. The Hague: Mouton.

Shimoda, K., Argyle, M. and Ricci Bitti, P. (1978). The intercultural recognition of emotional expressions by three national groups – English, Italian and Japanese. *Eur. J. Soc. Psychol.* 8, 169–79.

Taft, R. (1977). Coping with unfamiliar cultures. *In* Warren, N. (ed.). *Studies in Cross-Cultural Psychology*. London: Academic Press, vol. 1, pp. 121–53.

Taylor, D. M. and Simard, L. M. (1975). Social interaction in a bilingual setting. *Can. Psychol. Rev.* 16, 240–54.

Triandis, H. (1971). Work and leisure in cross-cultural perspective. *In* Abelson, R. P. *et al.* (eds). *Theories of Cognitive Consistency: a Sourcebook*. Chicago: Rand McNally, pp. 723–30.

Triandis, H. (1972). *The Analysis of Subjective Culture*. New York: Wiley.

Triandis, H., Malpass, R. S. and Davidson, A. R. (1972). Cross-Cultural psychology. *Biennial Rev. Anthropol.* 24, 1–84.

Triandis, H. C., Vassiliou, V. and Nassiakou, M. (1968). Three crosscultural studies of subjective culture. *J. Pers. Soc. Psychol.* 8, Monog. Suppl., part 2, pp. 1–42.

Tzeng, O. C. S. and Landis, D. (1979). A multidimensional scaling methodology for crosscultural research in communication. *In* Asante, M. K., Newmark, E. and Blake, C. A. (eds). *Handbook of Intercultural Communication*. Beverly Hills and London: Sage, pp. 283–317.

Watson, O. M. (1970). *Proxemic Behavior: A Cross-Cultural Study*. The Hague: Mouton.

Watson, O. M. and Graves, T. D. (1966). Quantitative research in proxemic behavior. *Am. Anthropol.* 68, 971–85.

Wujastyk, D. (1980). Causing a scandal in Poona. *The Times*, 24 April, p. 14.

Zigler, E. and Child, I. L. (1969). Socialization. *In* Lindzey, G. and Aronson, E. (eds). *The Handbook of Social Psychology*. Reading, Mass.: Addison-Wesley, vol. 3, pp. 450–589.

8 Methods of social skills training

MICHAEL ARGYLE

Introduction

Each of the chapters in this book has made some reference to the methods of social skills training (SST) which are being used for the skills described. In this chapter I shall consider these methods in more detail, and in particular discuss the evidence for the effectiveness of each method.

The traditional means of training, or lack of it, was training on the job; that is, simply learning by experience. This came to be supplemented in some cases, like teaching, by assigning an experienced performer of the skill to watch the trainee in action and give him helpful advice. Research into certain skills, like supervising work groups, led to the establishment of a body of knowledge about these skills which was passed on in the 1950s by educational methods, especially lectures and discussions. Developments in the social psychology of groups and the rise of the T-group and encounter-group movements led to the introduction of various kinds of group sensitivity training, especially for managers. Meanwhile role playing had been increasing in popularity and was boosted by the availability of videotape-recorders. It became widely used in assertiveness training and in SST for mental patients. The latest developments in SST are extensions of role playing based on new findings in social psychology – for example, in non-verbal (NV) communication, sequence analysis, and the study of social situations.

Not all of these methods have been found to be successful, and it is important to carry out follow-up studies. In a final section I shall discuss the design of such studies.

Learning on the job

Improvement of social skills by experience occurs mainly as a result of trial-and-error processes, together with the development of larger units of response. Unfortunately this seems to be a very unreliable form of training. A person can do a job for years and never discover the right social skills: some experienced interviewers have great difficulty with candidates who will not talk, for example. Or people can somehow learn the wrong skills by experience: Fiedler (1970) found that the more years of experience industrial supervisors had the *less* effective they were. Argyle, *et al.* (1958) found that supervisors often learnt the *wrong* things by experience, e.g. to use close, punitive and authoritarian styles of supervision.

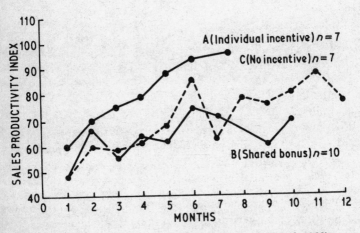

Figure 8.1　Learning curves for selling (from Argyle, *et al.* 1968).

Mary Lydall, Mansur Lalljee and myself conducted several studies of the learning of social skills on the job. In one of them an attempt was made to plot the learning curve for selling: this task was chosen because there is an objective criterion of success. Annual fluctuations in trade were overcome by expressing the sales of a beginner as a percentage of the average sales of three experienced sellers in the same department. It was found that there was an overall improvement, especially where there was an individual incentive scheme (Fig. 8.1). However, individuals responded in a variety of ways and while on average most improved, some did not and others got steadily worse.

Again it seems that simply doing the job doesn't always lead to improvement.

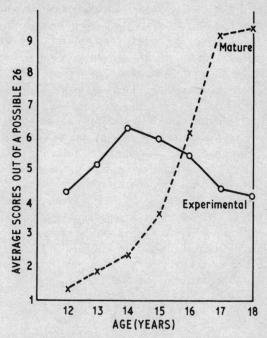

Figure 8.2   Experimental and mature solutions to social problems (from McPhail 1967).

McPhail (1967) studied the process of acquiring social competence during adolescence. He gave problem situations to 100 males and 100 females aged 12–18. Alternative solutions were offered and those chosen were found to change with age in an interesting way (Fig. 8.2). The younger subjects gave a lot of rather crude, aggressive, dominating responses. McPhail classified these as 'experimental' attempts to acquire by trial and error social skills for dealing with the new situations that adolescents face. The older ones, on the other hand, chose more skilful, sophisticated social techniques, similar to those used by adults.

Learning on the job has a great advantage over all other forms of training: there is no problem of transfer from training to real life. Very often an experienced practitioner is assigned to help the trainee by observing him in action and offering helpful advice. Several studies show that for this kind of coaching and feedback to work, certain conditions must be met:

(1) Clear feedback must be given on what the trainee is doing wrong. Gage, *et al.* (1960) asked 3900 schoolchildren to fill in rating scales to describe their ideal teacher and how their actual teachers behaved; the results were shown to half of the teachers, who subsequently improved on ten of the twelve scales, compared with the no-feedback group. This kind of feedback is usually not available.

(2) To improve performance new social techniques must be found. The best way of generating new responses is for an expert to suggest them and to demonstrate them. Learning on the job can be speeded up by imitating successful performers, but it may not be clear exactly what they are doing.

(3) Learning on the job will occur if there is a trainer on the spot who often sees the trainee in action and holds regular feedback and coaching sessions. The trainer should be an expert performer of the skill himself and should be sensitized to the elements and processes of social interaction. The success of such coaching will depend on there being a good relation between the trainee and the supervisor.

Educational (intellectual) methods

For all social skills there is a certain amount of new information and understanding to be acquired. Early attempts to teach, for example, supervisory skills by lecture and discussion were found to have very little effect, so that educational methods were generally abandoned. The motor-skill model suggests that purely intellectual methods would not be much good — you can't learn to swim by reading books about it. However, in cross-cultural training successful use has been made of educational methods alone or in combination with other techniques (pp. 188 ff.). This is an area where there is a lot to be learnt about the other culture, its rules, ideas and values.

Reading

This is one of the most widely used methods of education. Many self-improvement books in different areas have been produced, particularly in the USA, such as that by Carnegie (1936) which has been widely read. More recently there have been a number of books on do-it-yourself assertiveness training, e.g. Bower and Bower (1976), though no follow-up studies have been carried out. 'Bibliotherapy' produces short-term benefits at least in a number of areas, including weight reduction, study behaviour, fear reduction and exercise, though often some minimal therapist contact is needed (Glasgow and Thelen 1978). One area of social skills where reading is regularly used is for inter-cultural communication – where, for example, pro- grammed texts based on critical incident surveys in the form of Culture Assimilators have been quite successful (Fiedler, *et al.* 1971, and see pp. 188 ff.). However, in this field it is normal to combine learning by reading with more active forms of training, and self-help manuals usually include guidance on exercises to be carried out. It may be necessary, however, to have someone who can take the role of trainer and give some feedback on performance.

Reading is potentially a very important form of SST, since it is possible to reach many people.

Lectures

Lectures can be given in which various aspects of skill are explained, followed by discussion. They may focus on the basic principles of social behaviour or on the details of recommended social techniques. The lectures may be followed by discussion among the trainees, or there may be guided discussions without lectures. Experience with management training shows that lectures on 'human relations' are often very popular, and are a good means of conveying knowledge, though not a good way of changing attitudes. Can social skills be taught by means of lectures? Follow-up studies show that lectures on human relations lead to improved scores on questionnaires, but it has not been shown that any behavioural changes in skill are produced. There are certain difficulties about lectures. They are no good unless the audience is really interested in what the lecturer has to say, or unless he can make them interested by the forcefulness of his presentation, and unless he has a manner and status which make him personally acceptable.

There is some evidence that *group discussions* on human rela- tions problems, without lectures, lead to changes in questionnaire measures. Group discussion can also be looked at as a kind of role playing of group problem solving; it has been found that it results in improved committee skills (Maier 1953). One problem about group methods is that there may be little input of new information to the group. There can be various degrees of feedback, instruction or other guidance from the trainer; probably the more there is the more successful the results. In so far as the method works, success is partly due to the better assimilation of material put over in lectures or otherwise and through practice in group behaviour. It is also possible for a skilful leader to bring about changes in attitudes and values as a result of discussion.

The case-study method

This has been used for teaching social skills to schoolchildren. A case is presented by the teacher, or from a textbook, which illustrates problems such as dealing with authority, emotional problems at home or moral dilemmas. The case is then discussed by the class under the guidance of the teacher (McPhail 1972).

I shall refer to the use of films for modelling in connection with role playing. A number of suitable films are now available, mainly for management skills. Social skills trainers often make up their own videotapes for modelling behaviour for trainees. However, no follow-up studies are yet available on the use of films for this purpose. Films have been used for training in manual skills for some time, and these are found to be successful under certain conditions: if the learner has to try out part of the skill after each piece of film, if there is discussion before or after the film, if the film is shot from his point of view − e.g. over his shoulder − and if appropriate use is made of slow motion, animation and sequences of stills, showing the successive steps in the skill. Again, it looks as if films can play an important part in an overall training scheme but are not much use alone; and so far there are very few suitable films available.

However, there is some evidence that the combination of these methods, especially of lectures and group methods, can lead to increases in social skill. Sorensen (1958) studied 205 managers before and after such a course, and 267 controls. Ratings by other managers showed that the course trainees were rated as more co-operative, self-confident, poised and higher on 'consideration' (i.e. looking after the welfare of subordinates) and delegation. It is quite likely that neither lectures, reading, nor group discussion alone will do the job, but that together they can be very successful, especially in conveying knowledge about effective skills, understanding of social interaction and general awareness of interpersonal phenomena. On the other hand, managers may be better able than the other people to apply on the job new principles that they have learnt.

Training in groups

Therapy groups, T-groups and encounter groups were designed for different forms of training. The only kind of group training which has been recommended by a contributor to this book is the T-group (Chapter 5).

T-groups

The members of a T-group spend their time studying the group and the processes of social interaction that take place in it. The trainer typically starts the group off by saying 'My name is _____, and I am the appointed staff trainer of this group and am here to help you in the study of this group as best I can'. T-groups consist of about twelve trainees who meet for a series of 2-hour periods, weekly or during a residential course. The T-group sessions are often combined with lectures, role playing and other activities but the T-group is regarded as central.

The topic is unusual: one of the basic rules of T-groups is that conversation must be confined to the 'here and now', i.e. what is happening in the group. The role of the trainer is unusual: he does not take charge or act as a leader, but occasionally intervenes to make interpretations, provide feedback or draw attention to particular problems. He behaves rather like a group therapist, except that he does not discuss the personalities or problems of individuals, but rather the common problems of the group. This abdication of the role of the leader causes some perplexity and annoyance at first: the members are unfamiliar with the situation and seek help from the trainer in coping with it. During the early sessions there is a struggle for dominance, and questions of intimacy and friendship are sorted out. For females in particular it is found that there are problems of dependency − who to be dependent on, and how dependent to be.

Behaviour in T-groups has a strange quality, the conversation is somewhat stilted and embarrassed, and some members either do not take part at all, or couples engage in irrelevant conversation ('pairing'). A curious pattern of role differentiation has been reported in Harvard T-groups, including 'distressed females', who take little part in the conversation, 'paranoid' and 'moralistic' resisters, who oppose the official task and 'sexual scapegoats' who present their masculinity problems for the group to study (Mann, *et al.* 1967).

One of the main things that the trainer does is to teach people to give and receive feedback, so that members may become aware of the effect of their behaviour on others and find out how others see them. The trainer shows how to make non-evaluative comments on the behaviour of others, and tries to reduce the defensiveness of those whose behaviour is being commented upon. Feedback is provided in other ways; members may take turns to act as observers who later report back to the group, tape-recordings of previous sessions are studied and analyses by professional observers may be presented.

Encounter groups

These groups use a number of exercises designed to give experience of intimacy and other social relationships. Here is an example of the exercises used at Esalen (Schutz 1967).

To help people who have difficulty in giving or receiving affection or who avoid emotional closeness:

(1) 'Give and take affection'. One person stands in the centre of a circle with his eyes shut; the others approach him and express their feelings towards him non-verbally however they wish — usually by hugging, stroking, massaging, lifting, etc.
(2) 'Roll and rock'. One persons stands in the centre of a circle, relaxed and with his eyes shut; the group pass him round the group from person to person, taking his weight. The group then picks him up and sways him gently backwards and forwards, very quietly.

How successful are T-groups and encounter groups? This is a matter on which there is considerable disagreement. Those who have been in these groups report that they have been through a powerful experience. Smith (1975) reviewed thirty-one controlled follow-up studies, of which twenty-one showed positive results. The effect was greatest on self-reports, rather less on ratings by others or on organizational effectiveness. Most studies have found that the majority are unchanged. The disagreement arises over how many become *worse*. This issue has not yet been resolved. Several studies show a 'casualty rate' of 8 per cent, including the rather careful study by Lieberman, *et al.* (1973), and a number of those reviewed by Hartley, *et al.* (1976). Other studies suggest a lower rate of 1–2 per cent, and this is the figure favoured by Georgiades and Orlans in Chapter 5. It is agreed, however, that the rate of disturbance is highest with certain kinds of trainer, especially those who encourage confrontation and the expression of anger.

Some people may go to groups as a last resort before seeing a psychiatrist, some may become worse as a result of other changes in their lives. It has been said in defence of T-groups that they may be the only way of dealing with particularly difficult members of staff – who are reformed or never seen again. A problem with encounter groups which use a lot of bodily contact is that some clients end up with a different spouse from the one they started with. Some people enjoy the groups so much that they lose interest in ordinary life and want to spend all of their time having 'deep and meaningful experiences' in groups.

In a later section I shall discuss the need to accompany SST with various aspects of personal growth, such as increased self-confidence, reduced anxiety and new ideas and principles. Part of the attraction of group methods is that they provide opportunity for group discussion, self-disclosure of anxiety and thrashing out general problems connected with the job. However, I do not believe that the special procedures used in T-groups or encounter groups are what is needed. It might be better to have a series of group discussions led by a sympathetic and experienced leader who is familiar with the problems of young doctors, managers, etc., and who can help them make the necessary emotional and cognitive changes.

Role playing

Most forms of SST are varieties of role playing. Role playing consists of trying out a social skill away from the real situation in the laboratory, clinic or training centre on other trainees or role partners provided for the purpose. The training usually consists of a series of sessions which may last from 1 to 3 hours, depending on the size and stamina of the group. In each session a particular aspect of the skill or a particular range of problem situations is dealt with. There are three main phases to role-playing exercises:

(1) There is a lecture, discussion, demonstration, tape-recording or a film about a particular aspect of the skill. This is particularly important when an unfamiliar skill is being taught or when rather subtle social techniques are used. The demonstration is important: it is known as 'modelling'.

(2) A problem situation is defined and stooges are produced for trainees to role play with for 7–15 minutes each. The background to the situation may be filled in with written materials, such as the application forms of candidates for interview, or background information about personnel problems – the stooges may be carefully trained beforehand to provide various problems, such as talking too much or having elaborate but plausible excuses. It is found in microteaching that it is better to use real pupils than other trainee teachers as stooges, although this is a lot more trouble to arrange; the same probably applies to other areas of SST.

(3) There is a feedback session, consisting of verbal comments by the trainer, discussion with the other trainees, and playback of audio- or videotapes. Verbal feedback is used to draw attention, constructively and tactfully, to what the trainee was doing wrong and to suggest alternative styles of behaviour. The tape-recordings provide clear evidence for the accuracy of what is being said.

(4) There is often a fourth phase, in which the role-playing phase is repeated. In microteaching this is known as 're-teaching'.

There are a number of practical difficulties in conducting role playing, which are discussed in the chapter on social work skills in Argyle (1981, pp. 109–33). Trainers have to be given the right attitude to the training: it should be enjoyable and carried out sufficiently light-heartedly for them not to be embarrassed, but it should also be seen as a serious exercise, and they should not 'ham it up'. They may also need to be introduced gradually to the idea of videotaping.

Follow-up studies have found that role playing combined with coaching, feedback and modelling is successful with many kinds of professional clients and mental patients, and that it is one of the most successful forms of SST for all of these groups.

Feedback

This is one of the crucial components. It can be provided in several ways. It may come from the other members of the group, either in free discussion, discussion in smaller groups, questionnaires, or behavioural checklists. This must be done carefully or it will be disturbing to the recipients of the feedback; on the other hand, it is probably a valuable part of the training process for those observing.

It may be given by the trainer, who should be in a position to give expert guidance on the social techniques which are effective, and who may be able to increase sensitivity to the subtler nuances of interaction. He may correct errors – such as interrupting, looking or sounding unfriendly. He can suggest alternative social techniques, such as ways of dealing with awkward clients or situations. This has to be done very carefully: the trainer's remarks should be gentle and kind enough not to upset, but firm and clear enough to have some effect.

Audiotape-recordings may be taken and played back to the trainee immediately after his performance. I have found that the trainer's comments should precede the playback so that trainees know what to look for.

Videotape-recordings can be used in a similar way: a television film is played back to the trainee after his performance. This directs the trainee's attention to the behavioural (facial, bodily and gestural) aspects of his performance as well as to the auditory. It may be useful to play back the sound-tape separately to focus attention on sound.

It is usual for trainers to be generally encouraging and also rewarding for specific aspects of behaviour, though there is little experimental evidence for the value of such reinforcement. It is common to combine role playing with modelling and video playback, both of which are discussed below.

Modelling

Modelling can consist of demonstrations by one of the therapists, or the showing of films or videotapes. It is used when it is difficult to teach the patient by verbal description alone. This applies to complex skills for neurotic or volunteer clients, and to simpler skills for more disturbed patients and children. It is generally used in conjunction with role playing between role-play sessions and is accompanied by verbal instructions, i.e. coaching.

The success of modelling has been found to vary between different groups. For children it is the best technique (e.g. O'Connor 1972). For inpatients it is an essential part of SST — though the effectiveness of SST for these patients is in some doubt (Argyle 1981, p. 176). On the other hand, a number of outpatient studies have obtained good results without modelling (e.g. Argyle, *et al.* 1974), while McFall and Twentyman (1973) found that modelling made no contribution to assertiveness training for students, perhaps because they knew what to do already.

Modelling has been found to be most effective when the model is similar to the trainees, e.g. in age; when the model shows 'coping' rather than 'mastery' (i.e. is not too expert); when there is a verbal narrative, labelling the model's behaviour; when the model is warm and preferably live; with multiple models; and when the model's behaviour is seen to lead to favourable consequences (Thelen, *et al.* 1979).

Video playback

This is widely used in conjunction with role playing. Bailey and Sowder (1970) and Griffiths (1974) have reviewed some of the studies comparing the effectiveness of role playing with and without video, and came to the conclusion that there is little evidence for its usefulness. In fact nearly all the studies cited showed that patients did better when video playback was used. A later, carefully designed study on people who replied to an advertisement for SST obtained clearly better results with video; the criterion measure consisted of blind ratings of role playing. Video playback may not be equally suitable for all kinds of patients. Sarason and Ganzer (1973) believe that it should not be used with extremely anxious or self-conscious patients. Most people find it mildly disturbing at first, and that it increases self-consciousness; however this wears off by the second or third session. The author has found it particularly useful in training NV behaviour.

'Homework'

How do patients apply what they have learned in the training setting to real life? The main solution so far has been via 'homework'. Trainees are asked to try out the skills which they have just learnt several times before the next weekly session and to report back on any difficulties. They may be given written notes on the exercises to be carried out, and the steps in each, and they may be asked to keep notes of what happened. Falloon, *et al.* (1977) found that outpatients who were given structured homework assignments did better on nearly all outcome measures. There is usually difficulty in persuading patients to do the homework, however. This can be done by the use of token rewards, or by refusing to continue treatment until homework assignments have been completed.

Lindsay (1980) increased the generalization of conversational skills in schizophrenics, from clinic to ward, by using token rewards for homework. Matson, *et al.* (1980) selected for training those target behaviours which the nursing staff were willing to reinforce on the ward, such as talking clearly, being cheerful, complying with requests and appropriate NV communication.

These and other studies show that homework is a useful way of producing generalization.

Individual or group methods?

Most therapists and other trainers take patients in groups of 6–12, since groups have the advantage of providing a ready-made social situation and a series of role partners, make more economical use of therapist time and enable patients to feel more at ease in the company of other similar people. On the other hand, groups of patients contain bad models, so there should be more than one trainer present, and it is difficult to concentrate on individual problems. One solution is to start patients in groups but also to give them some individual sessions for particular problems. Very disturbed or regressed patients should be taken individually (Trower, *et al.* 1978). There is no problem in using groups for professional SST.

Interviewer training

One of the first social skills to be taught by role playing was selection interviewing. In the course for selection interviewing devised by Sidney and Argyle (1969) some of the exercises were designed to teach participants how to deal with 'awkward' candidates (see above pp. 17 ff.). Trainees interview trained stooges who talk too much, too little, are nervous, bombastic, anxious and so on. Each role-playing session on this course begins with a lecture and a film about the problems to be role played. There is also training in how to assess

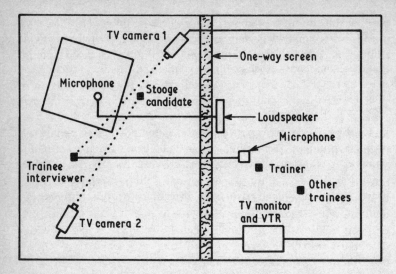

Figure 8.3 Laboratory arrangements for interviewer training.
VTR = videotape recorder.

stability, judgment, achievement, motivation, etc. in the interview and how to avoid common errors of person perception.

Role playing can be conducted without the use of any specialized equipment but it is greatly assisted if certain laboratory arrangements are available. An ideal set-up for interviewer training is shown in Figure 8.3. The role playing takes place on one side of a one-way screen and is observed by the trainer and other trainees. A videotape is taken of the role playing. The trainer is able to communicate with the role player through an ear-microphone; the trainer can give comments and suggestions to the trainee while the role playing is proceeding. (I once had to advise an interviewer trainee dealing with an over-amorous 'candidate' to move his seat back 3 feet.)

Microteaching

This is now widely used for training teachers. A trainee teacher prepares a short lesson, and teaches five or six children for 10–15 minutes; this is followed by a videotape playback and comments by the trainer, after which the trainee teaches the same lesson again. There are usually a number of sessions, each being devoted to one particular teaching skill: asking higher-order questions, encouraging pupil participation, explaining clearly with examples, etc. This form of training is found to be very much faster than alternative forms of training and is probably the best way of eliminating bad teaching habits (Brown 1975; Peck and Tucker 1973).

Management skills

Rackham and Morgan (1977) have developed a set of procedures for training in committee work, chairmanship, selling and related skills. They use a set of categories, which is modified for particular skills, containing items such as content proposals, procedural proposals, building, supporting, disagreeing, defending/attacking, testing understanding, summarizing, seeking information, giving information (the Chairman list). Good and bad performers at the skill are nominated and their rates of using the categories compared. Trainees learn the use of the categories and participate in role-play exercises while observers record how often they use the categories. The trainer then gives feedback, consisting of information about each trainee's score on the categories, which are compared with the rates for the good performers. Follow-up studies have shown positive results, though these studies were not very carefully controlled.

Special training methods

Some forms of SST have drawn heavily on the results of research in social psychology, and this is particularly the case with the Oxford form of SST. It has been found that facial expression, tone of voice, and other aspects of NVC make a major contribution to social competence, and that trainees often need to improve their performance in this sphere. Here are some examples, related to the processes described in Argyle (1981, Chapter 1).

Expression of NV signals

In training mental patients it is common to coach them in NV communication, since they are often very inexpressive in this sphere, or send NV messages which are hostile rather than friendly. Study of a patient's role-played performance in the clinic shows which NV signals require training, though the commonest ones are face and voice. Facial expression can be trained with the help of a mirror and later with a videotape-recorder. Trainees are asked to take part in short conversations, while expressing certain emotions in the face: sad, happy, etc. If there is difficulty in producing the correct expression in all parts of the face the photographs in Ekman and Friesen (1975) can be used as models. The voice can be trained with the help of an audiotape-recorder; trainees are asked to read passages from the paper in friendly, dominant, surprised, etc. tones of voice, and these are then played back and discussed. Details of correct vocal qualities for these emotions are given in Scherer (1979). I have found that neurotic patients improve rapidly with training. In other kinds of training there are always a few clients who need some help in this sphere. In most skills it is important to indicate attitudes such as warmth or concern, which requires the right NV signals.

Perceptual training

It may also be necessary to train people in the perception of NV signals. Some convicts, for example, can't tell when people are becoming upset or not, so that fights start. For professional skills like social work and psychotherapy it is important to be able to judge the emotional states of others. The Ekman and Friesen (1975) photographs can be used to train people to decode facial expression. Trainees can be taught to decode tones of voice by listening to tape-recordings of neutral messages produced in different emotional states. (Davitz 1964). In each case it is easy to test the subject, for example by finding out the percentage of recordings which they can decode correctly.

Jecker, *et al.* (1965) succeeded in training teachers to perceive more accurately whether or not pupils had understood what they were being taught. The measure consisted of a series of one-minute films showing children being taught; a subsequent question to the child (not shown in the film) found out if he really had understood or not. By showing teachers these films, and by drawing attention to the behavioural cues for comprehension, it was possible to increase their scores on a new set of similar films. A number of studies of teacher training have found that training in observation skills and practice with the Flanders category scheme improves teaching skills (Peck and Tucker 1973). This is probably because teachers become sensitized to the importance of pupil participation and the need to ask questions; they would already be able to perform the skilled moves which were needed.

Planning and the use of feedback

The social skills model (in Argyle 1981, p. 2) suggests some further points at which training can be useful. A common problem with mental patients is a failure to pursue persistent goals and a tendency to react passively to others. Assertiveness training is also directed at making the trainees take more initiative and pursue their goals. We have used special exercises for this problem; trainees are asked to carry out a simple skill, like interviewing, which requires that they take the initiative throughout. They can plan the encounter and take notes; the trainer communicates with an ear-microphone during the role playing if the performer runs out of conversation.

The social skill model also emphasizes the importance of responding to feedback. The social survey interview is an example (Chapter): if the respondent does not produce the desired information, revised questions are asked until this information is produced. The selection interview provides another example: the interviewer has to use different styles of behaviour with different interviewees. If a candidate does not talk enough the interviewer asks more open-ended questions, waits for the answers, and gives positive reinforcement for whatever is said. If the candidate talks too much, is very nervous, boastful, etc., other techniques are required.

Conversational sequences

All social skills use conversation, i.e. a sequence of utterances, and the control of sequences is an important part of the skill. Some neurotic patients are virtually incapable of sustaining a conversation at all (Argyle 1981, p. 166). Teachers need to be able to control such cycles of interaction as teacher lectures → teacher asks question → pupil replies (Argyle 1981, p. 17), and other longer cycles. During such repeated sequences there is also a build-up in the complexity of the topic being taught. The selection interview is similar: there is a certain structure of questions–answers–modified and follow-up questions, and a structure of episodes and sub-episodes, based on topics and sub-topics (Chapter 1). Every social skill uses certain conversational sequences, which can be learned. And every social skill has a number of difficulties in this sphere, for which the solution can be taught. Salesmen may have difficulty in controlling interaction with the client; doctors may find it is hard to terminate encounters; survey interviewers may have to deal with respondents who wander off the point.

Situational analysis

Some mental patients and many ordinary adults have difficulty in coping with specific social situations. In a number of professional social skills the performer has to deal with a variety of situations, as in the cases of social workers and supervisors. It would be possible to include in the training some analysis of the main situations involved, and especially of those which are found difficult, in terms of goals and goal structure, rules, roles, etc. (Argyle 1981, pp. 174–5). Situational analysis has been used in the treatment of obesity, by discovering the situations in which over-eating occurs, and the series of events leading up to it (Ferguson 1975). A similar approach is commonly used in the treatment of alcoholics.

Taking the role of the other

Chandler (1973) succeeded in improving ability to see the other's point of view (and reducing delinquency) by means of exercises in which groups of five young delinquents developed and made video recordings of skits about relevant real-life situations, in which each member of the group played a part.

Self-presentation

In addition to the usual role-playing exercises, trainees can be given advice over clothes, hair and other aspects of appearance. Their voices can be trained to produce a more appropriate accent or tone of voice. There is a correlation between physical attractiveness and mental health, and some therapeutic success has been obtained by improving the appearance of patients. The recidivism of male criminals has been reduced by removing tattooing and scars.

Some problems of social skills training

Prerequisites for successful training

To carry out role-playing or most other forms of training it is necessary to have a working vocabulary of the main elements in the repertoire for the skill being taught. This is needed to communicate with trainees, to label their behaviour, and for them to monitor their own performance. A vocabulary is needed in the following areas:

(1) Speech acts, e.g. makes interpretations, reflects feelings.
(2) Speech contents, e.g. candidate's exam results, leisure activities, future plans.
(3) NV communication, e.g. smiles, nods head.
(4) Actions, e.g. takes temperature, looks at tongue.

The same NV signals are used in all skills, though some will be more common in a particular case. Bales (1950) produced what was intended to be a universal set of twelve speech acts, but later work in particular fields has led to the development of categories tailored to the special needs of teaching, psychotherapy, etc. In some fields, however, there are a number of different category schemes, reflecting the ideas and interests of different investigators or trainers. A very large number of systems are used for training teachers, for example (Simon and Boyer 1970).

A second prerequisite is knowledge of the skills which are most successful. Without such knowledge the trainer has to fall back on common sense, which is very fallible in this field. Each of the chapters in this book has indicated the extent of research on the optimum skills, and it is clearly less extensive in some areas than others. A serious problem is that the most effective social skill may vary with situational factors such as characteristics of the others being handled; this has been studied most in the case of supervisory skills. The best skills may also be different in different classes, races or other sub-cultures. This is a particular problem with training mental patients, since the skills which they should be taught may vary considerably with their social class, etc.; and very often this kind of knowledge is not yet available.

Transfer to real life

Training on the job is the only form of training which does not have to face the problem of transferring the skills which have been learnt in training centre, lab or clinic to the real world. T-groups are very different from the outside world and it has been said that all they train for is other T-groups. Role playing usually tries to deal with this problem by means of 'homework'; however, trainees are often reluctant to try the new skills out or may refuse to do so. Some role playing uses quite realistic simulations of the real situations; examples are the model villages constructed in the Caribbean and Hawaii for preparing members of the Peace Corps for Latin America and the Far East (Guthrie 1966). There are other ways of preparing trainees for varied and unexpected incidents in real life. More abstract principles of behaviour may be taught, as opposed to specific skills. An example is 'be rewarding', where this can take a variety of forms. Another is to learn to watch out for feedback and to take rapid corrective action. If trainees learn the basic principles of situational analysis they can be on the look out for the particular rules, roles and other features of totally new situations. It is also possible to acquire the habit of learning new social skills whenever these are needed — by imitation of successful performers, and studying the behaviour which is effective and ineffective.

There may be resistance to trying out new social skills because it is feared they will be found surprising by friends and colleagues and will be rejected. Sending managers off to a remote training course is often unsuccessful for this reason. The solution is to carry out training within the organization and to show that senior members are in support of the behaviour being advocated by using them as trainers in some way. Some forms of SST for patients use their friends or family in the training in the same way.

The search for more economical methods

If most professional people, many mental patients and perhaps 10 per cent of the normal population need, or could profit by, SST, who is going to administer it? The use of group methods is common, because it saves trainer time and provides role partners. What of do-it-yourself methods? We have seen that 'bibliotherapy' can be useful, if combined with more acted methods.

Some microteaching has been carried out without trainers: trainees simply record and play back videotapes of their own role playing and compare them with films of model teachers. However, microteaching is more successful when there is a trainer (Peck and Tucker 1973).

Another method is simply to arrange for clients to have practice encounters with one another without a trainer. Arkowitz has found that this is successful in improving subjective feelings of confidence and reducing anxiety after six practice dates (Arkowitz, *et al.* 1978) or twelve practice interactions with potential same-sex friends (Royce and Arkowitz 1978). However, although subjective feelings were improved, there was no observed improvement in social skill.

The need for other aspects of personal growth

For several skills it is not enough simply to learn the correct social moves, or it may be impossible to use these skills unless certain emotional problems have been dealt with. Some skill situations commonly arouse anxiety — for example public speaking, dealing with hostile people or opposing others. Learning effective skills is certainly part of the answer and practising the skill successfully helps but may not be sufficient. Performers may simply lack the self-confidence to take the responsibilities required, for example, by doctors and managers. This entails a change in self-image, which is usually produced by simply doing the job. Training courses often include an element of guided group discussion, which can help trainees to talk about these matters, and receive some social support. Newcomers to a social role often need some other set of ideas or principles which will help them to deal with the moral, political or wider (even philosophical) issues at stake: for example, doctors may be concerned about how long to keep patients alive, social workers about conflicts between the demands of law and the interests of clients. Again, this goes beyond social skills and can be tackled by guided group discussion, together with reflecting on the experience of performing the skill.

Methods of assessing training procedures

The assessment of social competence

Any assessment of the effectiveness of SST requires some means of measuring the social competence of individuals.

Objective assessment of competence   The best way of measuring social competence is by a performer's success in attaining the goals of the job or situation which he is in. The competence of a salesperson, for example, could be measured by their average level of sales over a period of time compared with others' selling in the same department. In practice there are often complications; in the sales cases there may be informal rules to the effect that the senior salesperson deals with customers first, or deals with more expensive items. There may be more than one index of success: for example, the supervisor of a working group is trying to maximize the quantity and quality of output, and to keep down absenteeism, labour turnover and accidents. If such indices have low or negative correlations with each other it is necessary to weight them or decide on cut-off points for acceptable performance. Again it may be difficult to compare different supervisors since there are other differences between their departments; for example, in the nature of the work and the characteristics of their working groups. Some of these difficulties are less important in follow-up studies, since these study *differences* before and after training in the same departments.

Assessing the competence of patients, or of otherwise normal adults or children, in everyday non-work settings is more difficult. It is easier to recognize failures of skill rather than success: the extent to which an individual cannot communicate clearly or effectively, cannot establish or maintain relationships, quarrels and annoys other people or finds social situations difficult and stressful. However, success can be assessed with self-ratings, or ratings by others.

Ratings   Many follow-up studies have used ratings by the supervisor, of those being trained or by colleagues. There may be ratings on a number of different aspects of performance, including competence in several sub-tasks or situations. Such ratings are more useful if the supervisor sees the trainee in action on several occasions, though this is not entirely satisfactory since the trainee may perform differently when watched. Raters may be subject to various kinds of error in the use of rating scales, such as halo effect and central tendency. Ratings are not very satisfactory for before-and-after studies, since the attitudes of raters towards the method of training may affect their willingness to indicate improvement or deterioration after training.

Subjective reports   These have been widely used in psychiatric SST, and are acceptable there since one of the goals of training is to modify subjective feelings of anxiety and discomfort. Self-reports of competence are much less satisfactory; trainees can learn the right answers without their behaviour being affected at all (Fleishman 1955). And asking people how much they enjoyed a training course is of very little value — they usually say they enjoyed it very much.

Role-played tests   I have said that role-played tests for assertiveness in mental patients have been found to have rather low validity (Argyle 1981, p. 162). For all role-played tests it is probably necessary to sample a number of tasks, and to ensure that these are similar to real life. Such tests have been used in a number of studies, and have the advantage over objective measures of providing exactly the same situation for all trainees. In a follow-up study the pre-test provides some practice, so that an untrained control group is essential.

Other measures

I have discussed some specialized forms of training which focus on particular aspects of performance, and hence may require measures of special aspects of social competence. Examples are accuracy of perception, level of anxiety, and voice quality. On the other hand, improving such specific aspects of social performance is no use unless it increases overall social competence.

The design of follow-up studies

The most obvious procedure is to compare the competence of trainees before and after training. The difficulty with this design is that with the passage of time other changes may affect the trainees and they may improve anyway through being more familiar with the test situation (for some kinds of measure). Many investigators have used the before-and-after study with control groups of similar but un-trained individuals. In practice it may be difficult to find a suitable control group. Why were the trainees being trained but not the members of the control group? Perhaps the trainees were in greater need of training. A design which avoids this is the after-only design, which compares people after a period of training, after different kinds of training or after no training. Members of the control group may be scheduled for training at a later date. This method, however, requires very careful matching of the groups of individuals to be compared and needs larger numbers.

Sometimes, it would be unacceptable to withhold training for a control group, for example with patients. One procedure here is the cross-over design in which two groups of trainees receive treatments A and B, or treatment and no treatment, in reverse order (Argyle, *et al.* 1974).

Another method is the multiple baseline procedure for follow-up of individuals in intensive clinical trials, which was described earlier.

References

Argyle, M. (ed.) (1981). *Social Skills and Health*. London: Methuen.

Argyle, M., Bryant, B. and Trower, P. (1974). Social skills training and psychotherapy: a comparative study. *Psychol. Med.* 4, 435–43.

Argyle, M., Gardner, G. and Cioffi, F. (1958). Supervisory methods related to productivity, absenteeism and labour turnover. *Hum. Rel.* 11, 23–45.

Argyle, M., Lalljee, M. G. and Lydall, M. (1968). Selling as a social skill (mimeo).

Arkowitz, H. (1977). Measurement and modification of minimal dating behavior. *In* Hersen, M. (ed.). *Progress in Behavior Modification 5*. New York: Academic Press.

Bailey, K. G. and Sowder, W. T. (1970). Audiotape and videotape self-confrontation in psychotherapy. *Psychol. Bull.* 74, 27–137.

Bales, R. F. (1950). *Interaction Process Analysis*. Cambridge, Mass.: Addison-Wesley.

Bower, S. A. and Bower, G. H. (1976). *Asserting yourself*. Reading, Mass.: Addison-Wesley.

Brown, G. (1975). *Microteaching*. London: Methuen.

Carnegie, D. (1936). *How to Win Friends and Influence People*. New York: Simon and Schuster.

Chandler, M. J. (1973). Egocentrism and anti-social behavior: the assessment and training of social perspective-taking skills. *Dev. Psychol.* 9, 326–32.

Davitz, J. R. (1964). *The Communication of Emotional Meaning*. New York: McGraw-Hill.

Ekman, P. and Friesen, W. V. (1975). *Unmasking the Face*. Englewood Cliffs: Prentice-Hall.

Falloon, I. R. H., Lindley, P., McDonald, R. and Marks, I. M. (1977). Social skills training of out-patient groups: a controlled study of rehearsal and homework. *Br. J. Psychiat.* 131, 599–609.

Ferguson, J. M. (1975). *Learning to Eat: Behavior Modification for Weight Control*. New York: Hawthorn Books.

Fiedler, F. E. (1970). Leadership experience and leader effectiveness – another hypothesis shot to hell. *Org. Behav. Hum. Perf.* 5, 1–14.

Fiedler, F. E., Mitchell, R. and Triandis, H. C. (1971). The culture assimilator: an approach to cross-cultural training, *J. Appl. Psychol.* 55, 95–102.

Fleishman, E. A. *et al.* (1955). *Leadership and Supervision in Industry*. New York: Columbia University Press.

Gage, N. L., Runkel, P. J. and Chatterjee, B. B. (1960). *Equilibrium Theory and Behavior Change: an Experiment in Feedback from Pupils to Teachers*. Urbana, Ill.: Bureau of Educational Research.

Glasgow, R. E. and Rosen, G. M. (1978). Behavioral bibliotherapy: a review of self-help behavior therapy manuals. *Psychol. Bull.* 85, 1–23.

Griffiths, R. D. P. (1974). Videotape feedback as a therapeutic technique: retrospect and prospect. *Behav. Res. Ther.* 12, 1–8.

Gudykunst, W. B., Hammer, M. R. and Wiseman, R. L. (1979). An analysis of an integrated approach to cross-cultural training. *Int. J. Intercultural Rel.* 1, 99–110.

Guthrie, G. M. (1966). Cultural preparation for the Philippines. *In* Textor, R. B. (ed.). *Cultural Frontiers of the Peace Corps*. Cambridge, Mass.: M.I.T. Press.

Hartley, D., Roback, H. B. and Abramowitz, S. I. (1976). Deterioration effects in encounter groups. *Am. Psychol.* 31, 247–55.

Jecker, J. D., Maccoby, N. and Breitrose, H. S. (1965). Improving accuracy in interpreting non-verbal cues of comprehension. *Psychol. Schools* 2, 239–44.

Lieberman, M. A., Yalom, I. D. and Miles, M. B. (1973). *Encounter Groups: First Facts*. New York: Basic Books.

Lindsay, W. R. (1980). The training and generalization of conversation behaviours in psychiatric in-patients: a controlled study employing multiple measures across settings. *Br. J. Soc. Clin. Psychol.* 19, 85–98.

Maier, N. R. F. (1953). An experimental test of the effect of training on discussion leadership. *Hum. Rel.* 6, 161–73.

Mann, R. D., *et al.* (1967). *Interpersonal Styles and Group Development*. New York: Wiley.

Matson, J. L., *et al.* (1980). A comparison of social skills training and contingent attention to improve behavioural deficits of chronic psychiatric patients. *Br. J. Soc. Clin. Psychol.* 19. 57–64.

McFall, R. M. and Twentyman, C. T. (1973). Four experiments on the relative contribution of rehearsal, modeling, and coaching to assertive training. *J. Abnorm. Psychol.* 81, 199–218.

McPhail, P. (1967). The development of social skill in adolescents, paper to British Psychological Society (unpublished), Department of Education/ University of Oxford.

McPhail, P. (1972). *Moral Education in Secondary Schools*. London: Longmans.

O'Connor, R. D. (1972). Relative efficacy of modeling, shaping, and the combined procedures for modification of social withdrawal. *J. Abnorm. Psychol.* 79, 327–34.

Peck, R. F. and Tucker, J. A. (1973). Research on teacher education. *In* Travers, R. M. W. (ed.). *Second Handbook of Research on Teaching*. Chicago: Rand McNally.

Rackham, N. and Morgan, T. (1977). *Behaviour Analysis and Training*. London: McGraw-Hill.

Royce, W. S. and Arkowitz, H. (1978). Multimodal evaluation of practice interactions as treatment for social isolation. *J. Consult. Clin. Psychol.* 46, 239–45.

Sarason, I. G. and Ganzer, V. J. (1971). *Modeling: an Approach to the Rehabilitation of Juvenile Offenders*. Baltimore, M.D.: U.S. Departments of Health, Education and Welfare.

Scherer, K. R. (1979). Nonlinguistic indicators of emotion and psychopathology. *In* Izard, C. E. (ed.). *Emotion in Personality and Psychopathology*. New York: Plenum, pp. 495–529.

Schutz, W. C. (1958). *FIRO: A Three-Dimensional Theory of Interpersonal Behavior*. New York: Holt, Rinehart and Winston.

Sidney, E. and Argyle, M. (1969). *Training in Selection Interviewing*. London: Mantra.

Simon, A. and Boyer, E. G. (eds) (1974). *Mirrors for Behavior, Classroom Interaction Newsletter*. Wyncote, Penn.: Communication Materials Center, 3rd edn.

Smith, P. B. (1975). Controlled studies of the outcome of sensitivity training. *Psychol. Bull.* 82, 597–622.

Sorensen, O. (1958). *The Observed Changes Enquiry*. New York: G.E.C.

Textor, R. B. (ed.) (1966). *Cultural Frontiers of the Peace Corps*. Cambridge, Mass.: M.I.T. Press.

Thelen, M. H., Fry, R. A., Fehrenback, P. A. and Frantschl, N. M. (1979). Therapeutic videotape and film modelling: a review. *Psychol. Bull.* 86, 701–20.

Trower, P., Bryant, B. and Argyle, M. (1978). *Social Skills and Mental Health*. London: Methuen.

Name index

Page numbers in italic type refer to the full reference given at the end of each chapter.

Subject index